Fast Reactions in Energetic Materials

Alexander S. Shteinberg

Fast Reactions in Energetic Materials

High-Temperature Decomposition of Rocket Propellants and Explosives

Translated from Russian by Maria Rusanova

Springer

Prof. Alexander S. Shteinberg
61 Fairlawn Drive
Berkeley CA 94708
USA
shteinberg@aol.com

Translated from Russian Edition (5-9221-0690-2), published by Nauka
in September 2006

ISBN: 978-3-540-78860-7 e-ISBN: 978-3-540-78861-4

Library of Congress Control Number: 2008932573

Cover design: WMXDesign GmbH

Printed on acid-free paper

9 8 7 6 5 4 3 2 1

springer.com

Preface for English Edition

This monograph is dedicated to the theoretical and experimental study of the macrokinetics and kinetics of the fast high-temperature decomposition of condensed energetic materials (explosives, pyrotechnical mixtures and gunpowders, as well as modern solid, liquid and hybrid rocket propellants). Numerous practical tasks related to the manufacture and application of the energetic materials can only be solved through the use of reliable information on the patterns, mechanisms and characteristics of their high-temperature decompositions (especially about their ignition and burning mechanisms).

In order to be able to evaluate the conditions needed for the safe use of rockets undergoing aerodynamic heating, the kinetic constants for the decompositions of the corresponding solid propellants are required. Liquid propellants are also affected by the intense heating that occurs in the nozzle cooling systems and precombustion chambers of rocket engines. Under certain conditions this can result in burning and explosions in the corresponding parts of the rocket engines. Therefore, data on the kinetics of the homogeneous and heterogeneous decompositions of liquid propellants (on the surfaces of the construction materials used) are extremely important for various applications in rocket and space engineering.

Kinetic constants of fast high-temperature decomposition of solid and hybride propellants and their components are of primary importance for theoretical analysis and practical applications. The extrapolation of low-temperature kinetic data into the high-temperature region gives temperature values for the burning surface that drastically disagree with experimental data obtained by thermocouple measurements. Since there are quite a few examples of such disagreements, the only solution is to develop methods (initially experimental) for direct measuring high rates of thermal decomposition and for estimating kinetic parameters at high temperatures. As explained in this book, classical isothermal methods of chemical kinetics cannot be used in this case.

The application of the new methods of nonisothermal kinetics to the study of fast decompositions of energetic materials at their ignition, burning and explosion temperatures is theoretically based. The kinetic parameters for the decompositions of polymers, ammonium perchlorate, nitrocellulose and others are calculated from experimental data on their linear pyrolysis or ignition (that is, under conditions of spatially nonisothermal heating). New patterns for the high-temperature decompositions of these energetic materials are described and characterized using the new methods, and kinetic constants for the initial stages are determined.

Approaches to modifying and optimizing DTA, TGA and DSC instruments and techniques so that they can be used to study the kinetics of fast reactions are proposed. Due to the original design of the experimental cells, the sample heating is spatially isothermal in nature and self-inflammation is excluded. These methods are used to study the reaction mechanisms and to determine rate constants for the fast high-temperature decompositions of a number of homogeneous and heterogeneous energetic materials, including nitrocellulose, ammonia copper chromate, ammonium perchlorate and ammonium perchlorate-based solid propellants.

In most cases, both the mechanism and the kinetic constants for the reactions associated with the high-temperature decomposition of energetic materials are different from those of low-temperature decomposition. It is important that the results from the theoretical analysis of the ignition and burning of the energetic materials studied, based on high-temperature experimental data, are in good agreement with experimental ignition delay periods and burning rates.

Some aspects of the high-temperature decomposition of so-called volatile energetic materials (including liquid nitrates, highly concentrated H_2O_2, hydrazine and ammonium perchlorate) are considered. Two simultaneous volumetric processes are characteristic of the macrokinetics of high-temperature conversions of these compounds: irreversible decomposition and phase transformations under the conditions of thermodynamic equilibrium at the interface (evaporation in liquids or dissociative sublimation in ammonium perchlorate). For volatile energetic materials, the formation of gas during their decomposition and thermal explosion is shown to occur in a different manner compared to nonvolatile energetic materials. Some instruments used to determine the kinetic parameters and critical conditions of thermal explosions of volatile energetic materials are described. Kinetic parameters of homogeneous and heterogeneous decomposition (on the surfaces of construction materials) of highly concentrated H_2O_2 and hydrazine are determined for wide temperature and reaction rate ranges. The correctness of the proposed thermal explosion theory for volatile energetic materials is confirmed by experimental data on the thermal explosion of highly concentrated H_2O_2.

Most of the work included here is published for the first time.

Berkeley, March 2008 *A.S. Shteinberg*

Foreword for Russian Edition

The kinetics of thermal decomposition reactions in condensed systems are currently studied with classical isothermal methods for measuring the rates of rather slow processes.

Modern experimental chemical physics methods used to study the kinetics of fast high-temperature gas-phase reactions are relatively well developed. They have been used in fundamental research for a long time, and the results from such studies have been successfully used for engineering applications.

In contrast, the kinetics of fast high-temperature reactions in condensed systems have not been studied until recently due to a lack of appropriate methods and instruments. This research area is of paramount importance to our understanding of a wide range of practically significant phenomena such as the combustion of polymers, explosives, blasting powders, and solid rocket propellants, decomposition, ablation, the erosion of heat shielding and construction materials under the impact of high-intensity heat fluxes, and the heating and decomposition of liquid propellants in rocket engine cooling systems.

In chemical kinetics a transition from one temperature region to another is often associated with a change in the limiting stage of the overall chemical process. Therefore, estimating the reaction rates at high temperatures by extrapolating kinetic data for the low-temperature region obtained via classical isothermal methods inevitably results in major errors. For instance, in the thermal destruction of a typical thermoplastic such as polymethylmethacrylate, extrapolating the low-temperature data yields the pyrolysis temperature of the burning polymer surface close to the temperature of the flame above the surface, which dramatically disagrees with the results of direct microthermocouple measurements. In contrast to gas-phase systems, the macrokinetics of the high-temperature transformations of many condensed materials are significantly affected by phase transformations (evaporation, sublimation). Due to low vapor pressures, these processes play a negligible role in the low-temperature region, but they become more important to the macrokinetics of the overall process as the temperature grows.

Thus, experimental determinations of kinetic and macrokinetic parameters for high-temperature condensed-phase reactions are absolutely indispensable if we are to gain deep insights into fundamental aspects of chemical kinetics, as well as for numerous important technical applications.

This book deals with the analysis of the kinetics of high-temperature decomposition of energetic materials under the conditions of spatially and temporally nonuniform temperature and concentration fields; in other words, when heat and mass transfer play significant roles. This area of process kinetics was termed "macrokinetics" by its founders D.A. Frank-Kamenetsky, Ya.B. Zeldovich and N.N. Semenov.

A.S. Shteinberg developed and tested a number of new methods and instruments for the quantitative study of the kinetics of fast high-temperature reactions in homogeneous and heterogeneous condensed systems. The extremely broad range of the rates measured by the author (covering 13 orders of magnitude, from slow decomposition rates under long-term storage conditions to the rates corresponding to ignition and combustion) is especially noteworthy and important for substantiating and generalizing the conclusions.

An important and unconditionally helpful feature of the book by A.S. Shteinberg (and one rarely encountered in scientific publications) is the way in which it combines theoretical and experimental parts, including descriptions of novel methods and instruments that were, for the most part, developed by the author. The book presents a large set of new data characterizing the high-temperature kinetics and thermal physics of many systems useful for technological applications, which means that the book can be considered a practical manual too.

A.S. Shteinberg is internationally recognized as a scientist with an extremely wide range of research interests, and an author of many experimental and theoretical studies in the field of the macrokinetics of the combustion and decomposition of homogeneous and heterogeneous condensed systems.

No similar books are available in the scientific literature, and this monograph will be helpful for undergraduates, graduate and postgraduate students, researchers, as well as engineers who have to deal with combustion and explosion phenomena and the high-temperature decomposition of condensed substances, materials, and structures in many aerospace, defense and civil applications.

Prof. A.A. Berlin
Full Member (Academician) of the Russian Academy of Sciences

Contents

Part II Study of the Macrokinetics of High-Temperature Decomposition of Condensed Materials Under Spatially Isothermal Heating

Lists of Abbreviations and Symbols

ACC	Ammonia copper chromate
AP	Ammonium perchlorate
PMMA	Polymethylmethacrylate
DINA	Dinitroxydiethylnitramine
AVD	Automatic volumetric device
DTA	Differential thermal analysis
TGA	Thermogravimetric analysis
DSC	Differential scanning calorimetry
TSI	Thermal stability index
h/c H_2O_2	Highly concentrated H_2O_2
A	Coefficient in the combustion law
a	Temperature diffusivity
Bi	Bio number
C	Specific heat capacity for inert compound
C	Specific heat capacity for reacting compound
D	Sample diameter
E	Activation energy
F	Surface area for a construction material (foil, film)
ΔH, L	Sublimation heat
k	Reaction rate constant
k_0	Preexponential factor in the Arrhenius law
M	Weight of inert compound
M	Michelson number
m	Weight of reacting compound
$\dot{m}$	Weight rate of linear pyrolysis (combustion)
n	Reaction order
P	Pressure
Q	Reaction thermal effect
$\dot{q}$	Heat evolution rate
q	Thermal flux rate
R	Gas constant

S	Heat exchange surface area in DTA- and thermal explosion-related expressions; surface area of the sample end in linear pyrolysis-related expressions
T	Temperature
t	Time
U, u	Rate of linear pyrolysis, combustion rate
W	Force pressing a sample to the heater in linear pyrolysis experiments
Z	Thickness of gas gap
α	Heat-exchange coefficient
β, γ, δ	Dimensionless parameters in the theory of thermal explosion
ε, μ	Correction factors taking block-calorimeter heat losses into account
λ	Thermal conductivity
ω	Linear heating rate
η	Conversion degree
$\dot{\eta}$	Conversion rate
ν	Index in the combustion law
Θ	Dimensionless temperature
τ	Dimensionless time
κ	Combustion constant in the Sreznevsky law
ρ, γ_{sol}	Density

Introduction

Modern energetic materials include explosives, blasting powders, pyrotechnic mixtures and rocket propellants [1, 2]. The study of high-temperature decomposition of condensed phases of propellants and their components (liquid, solid and hybrid) is currently of special importance for the development of space-system engineering [3, 4]. To better understand the burning mechanisms (stationary, nonstationary, unsteady) of composite solid propellants and their components, information about the macrokinetics of their high-temperature decomposition is required [5]. To be able to evaluate the ignition parameters and conditions of safe handling of heat-affected explosives, one needs to know the kinetic constants of their high-temperature decomposition. The development of new composite solid propellants characterized by high performance characteristics (high burning rates, high thermal stability, stability to intrachamber perturbations, and other aspects) is not possible without quantitative data on the high-temperature decomposition of composite solid propellants and their components [6].

The same reasons have resulted in significant theoretical and practical interest in the high-temperature decomposition of components of hybrid propellants. It is known that hybrid propellants have not been used very widely due to the low burning (pyrolysis) rates of the polymer blocks in the combustion chambers of hybrid rocket engines. To increase the burning rates it is necessary to obtain information about their relationships to the corresponding kinetic and thermophysical properties of the fuels. Thus, the study of high-temperature decomposition of composite solid and hybrid propellants is of primary importance due to the important role of decomposition-related processes in the ignition and burning of these fuels.

The study of the high-temperature decomposition of the condensed phase of modern liquid propellants is important due to different reasons than those mentioned above. The high-temperature processes that occur in the condensed phase during burning in liquid-propellant rocket engines do not play a very important role for most fuels (and oxidizers). However, the situation for processes that take place in intensively heated cooling systems and precombustion chamber nozzles of liquid-propellant rocket engines is completely different. This is particularly true of monopropellants (highly concentrated hydrogen peroxide and hydrazine), as well as some single-component organic nitrate liquid fuels (for example, propylene glycol dinitrate). The study of the macrokinetics of high-temperature decomposition and thermal explosion of monopropellants is very important due the ability of these materials to undergo spontaneous homogeneous and heterogeneous decomposition resulting in the release of a great deal of heat in liquid-propellant rocket engine systems. In addition, quantitative data on the kinetics of the high-temperature decomposition of solid and liquid propellants are required for the development of express methods of evaluating the thermal stabilities of rocket propellants.

When studying the kinetics of thermal decomposition at high temperatures, one normally considers relatively fast reactions to be characterized by a reaction time $t_r \approx [k(T)]^{-1}$, where k is the reaction rate constant and T is the characteristic temperature. It is obvious that this time should be considered in relation to some physical scale. In this case, the most reasonable scale is the characteristic transfer time [7]. In the case of non-diffusion-limited thermal decomposition, the characteristic sample preheat time ($t_p = d^2/a$, where a is the sample temperature diffusivity and d is the sample diameter) can be used as a reference time at low temperatures and reaction rates $t_p \ll t_r$. In this case, chemical reactions are spatially isothermal regardless of the sign and magnitude of the thermal effect. If the system is characterized by active heat transfer, the sample temperature is also independent of time.

When the temperature on the sample surface, T_S, is increased, or in the environment that exchanges the heat with the sample, the decomposition macrokinetics is characterized by a significantly different pattern. In this case, the T_S-related reaction time decreases and $t_p \approx \text{const} \neq f(T_S)$. At $t_p \approx t_r$ a thermal explosion takes place in the system along with an exothermic reaction. In the region above the thermal explosion initiation limit at $t_p \gg t_r$(long before the sample is preheated and decomposed to a noticeable degree) ignition takes place in the zone adjacent to the hot surface.

The theory of ignition and the transition from thermal explosion to ignition was first proposed by Zel'dovich [8] and then developed in greater detail by Merzhanov and coworkers. Processes of thermal explosion and especially ignition were shown to be characterized by a strongly pronounced nonisothermal spatiotemporal pattern. Because of this, none of the isothermal methods of classical chemical kinetics can be used to study high-temperature fast exothermic reactions. The state-of-the-art in this area was considered in our recent lecture at the 35th IPS [9].

In practice, anyone studying fast endothermal reactions (for example, processes accompanying the forced high-temperature destruction of polymers, such as the components of solid propellants or binders of ablative shields of space vehicles) faces similar significant difficulties. In this case, although the increase in temperature on the pyrolysis surface does not result in self-inflammation of the material, the temperature distribution within it is a function of time and spatial coordinates. Therefore, the attribution of the kinetics of the recorded thermal decomposition process to a certain temperature is very problematic. In this case, classical isothermal methods cannot be used for kinetic studies despite the absence of thermal explosion and ignition. However, even incomplete information (for example, a kinetic equation for the reaction cannot be obtained) about the rate constants of high-temperature reactions and the effective activation energies obtained from purely macrokinetic data is of significant interest and importance. The theory of nonisothermal methods of chemical kinetics originating from the classical works of Frank-Kamenetsky, Semenov, Zel'dovich and Merzhanov is based on this approach.

At the beginning of the studies considered in this book, the macrokinetic theory of linear pyrolysis did not exist and the experimental techniques were quite poor and could only be used within a very narrow range of parameters. In contrast, the ignition theory was quite well developed, but experimental ignition-based methods could only rarely be used to obtain data on high-temperature decomposition in condensed phases. Therefore, the kinetics of high-temperature decomposition of rocket propellant components and compositions were only poorly studied. Data on the high-temperature decomposition of ammonium perchlorate (AP, the most commonly used oxidizer in modern composite solid propellants) and model and commercial polymer fuels of composite solid and hybrid propellants were not available. Experiments on the linear pyrolysis of AP and polymers were carried out using faulty devices and techniques (for example, the heater temperature was considered to be identical to the temperature of the pyrolysis surface). A few data on the high-temperature decomposition of components of composite solid propellants could be found, but there was no information on their high-temperature (>240–300°C) decomposition kinetics. This situation can be explained by a lack of corresponding experimental methods. Thus, the author had to face many challenges connected with theoretical and experimental aspects of the study of the macrokinetics of high-temperature decomposition of homogeneous and heterogeneous components of rocket propellants under the conditions of linear pyrolysis, ignition and programmed heating as well as those related to designing, manufacturing and testing new devices (and methods).

In terms of the general differences between high- and low-temperature decomposition of condensed materials, one should keep in mind that the transfer of the system from one temperature region into the other can be accompanied by a change in the limiting kinetic stage. Therefore determining the rate constants at high temperatures by extrapolating from low-temperature data can result in very serious errors. Thus, the availability of experimental approaches for obtaining data on the kinetics of high-temperature fast decomposition of energetic materials in the condensed phase is of significant importance.

For volatile explosives and monopropellants (including various secondary explosives such as dinitroxydiethylnitramine, Tetryl and single-component propellants belonging to the class of liquid organic nitrates), as well as for highly concentrated hydrogen peroxide, hydrazine and AP, the decomposition macrokinetics at low and high temperatures are significantly different, even though the kinetic mechanism does not change. In this case, all of these materials that are different in nature and properties are characterized by a common feature: their ability to sublimate with heat absorption in the condensed phase. Due to this ability, decomposition takes place simultaneously in the condensed and vapor phases (the process rates generally differ markedly). At low temperatures the vapor pressure of, for example, H_2O_2 is low, the evaporation rate is also low, and the compound can be considered to be nonvolatile. However, as the temperature increases, the situation changes dramatically. Depending on the boiling point (determined by the ambient pressure and the vapor pressure as a function of temperature) and the rate constants of reactions in the condensed and gas phases at this temperature, the compound can exhibit nonvolatile explosive behavior or that of an almost inert liquid. The complex nature of these phenomena has resulted in some contradictions between the explanations of the mechanism for inflammation and thermal explosions of volatile explosives proposed by Belyaev ("the flash takes place in vapors") and by Merzhanov, Abramov and Barzykin ("critical conditions of thermal explosion are determined by kinetics of the condensed-phase reaction"). Since in rocket engines propellant decomposition can occur over a wide range of temperatures and pressures, the effect of volatility on decomposition (especially high-temperature) macrokinetics is of tremendous theoretical and practical interest.

This book is devoted to analyzing the state-of-the-art and further progress in the field of the theoretical and experimental study of high-temperature fast reactions in the condensed phases of energetic materials (including modern rocket propellants and their components as well as some explosives and homogeneous blasting powders). Comprehensive solutions for some important tasks are presented for the first time, since for many years these studies were "classified." The book is written in the form of a monograph, in which a number of ideas and aspects are considered and developed from a certain point of view. A review of the relevant available literature and an analysis of all research directions in this field were not among the goals of the author. Therefore the list of references only includes papers directly connected to the topic of the book.

Fundamental contributions in this research area were made by Summerfield, Williams, Beckstead, Brill, Kuo, Thompson, Waesche, Frank-Kamenetsky, Zel'dovich, Semenov, Belyaev, Merzhanov, Novozhilov, Manelis, Barenblatt, Berlin, Fortov, and others.

The book can be considered the result of long-term cooperation between the author and researchers from the Semenov Institute of Chemical Physics (Moscow, Russia). The author is especially grateful to A.G. Merzhanov, one of the founders of the scientific school of nonisothermal kinetics, for his support.

Important contributions to the design and manufacture of new devices for studying nonisothermal kinetics were made by Goncharov, Kochetov, Slutsker, Sokolova, and Shirokova. The author is grateful to Rusanova, Andreeva and Grigoryants for their help in preparing this book for publication.

References

1. Committee on Advanced Energetic Materials and Manufacturing Technologies (2004) Advanced energetic materials. Committee on Advanced Energetic Materials and Manufacturing Technologies, National Research Council of the National Academies, Washington, DC
2. Davenas A (1993) Solid propellant propulsion technology. Pergamon, Oxford
3. Beckstead MW (2000) In: 29th JANNAF Propellant Development & Characterisation Meeting, Cocoa Beach, FL, USA, 8–12 May 2000
4. Agrawal JP (1998) Prog Energy Combust Sci 24:1
5. Bakhman NN, Belyaev AF (1967) Combustion of heterogeneous condensed systems. Nauka, Moscow
6. Miller RS (1996) In: Brill TB, Russel TP, Tao WC, Wardle RB (eds) Decomposition, combustion, and detonation chemistry of energetic materials. Materials Research Society, Pittsburgh, PA, p 3

7. Frank-Kamenetsky DA (1969) Diffusion and heat transfer in chemical kinetics. Plenum, New York
8. Zel'dovich JaB (1939) Zh Eksp Teor Fiz 9:1530
9. Shteinberg AS, Berlin AA (2008) In: 35th IPS, Fort Collins, CO, USA, 13–18 July 2008

Part I
Study of Macrokinetics of High-Temperature Decomposition of Condensed Materials Under Spatially Nonisothermal Heating

Chapter 1
Theory of the Linear Pyrolysis of Condensed Materials

Abstract Linear pyrolysis is the one-dimensional propagation of the reaction front in thermally decomposing condensed compounds. Numerous energetic materials and their components (polymer binders and oxidizers of solid rocket propellants, homogeneous blasting powders, explosives and others) decompose via linear pyrolysis under intense heating and burning. A theory of linear pyrolysis including analytical expressions relating the rate of linear pyrolysis (burning) to kinetic constants for the decomposition that occurs in a thin layer of the compound near its hot surface is presented. The conclusions of this theory are of significant importance to our understanding of processes such as the burning of solid rocket propellants and explosives, the ablation of space vehicle heat shielding, etc.

1.1 Introduction

Linear pyrolysis is the steady one-dimensional propagation of the thermal decomposition front in a condensed substance under conditions where heat is supplied by an external source.

Schultz and Dekker [1] were the first to propose the use of this phenomenon to study the kinetics of the thermal decomposition of condensed materials. Recently, the interest of physical chemists in this approach has been growing [2, 3] (you can find a list of publications on linear pyrolysis in [4, 5]).

In the experiments, a sample is pressed onto a heated hard surface kept at a constant temperature, T_0. The velocity of the motion of the sample towards the hard surface, U (equal to the linear decomposition rate) is measured. This method is especially convenient for studying the thermal decomposition of gasifiable compounds.

If at an instantaneous temperature of the hot surface of the compound, T_S, its decomposition results in the formation of condensed products and it occurs with significant heat release, causing ignition (also termed "unsteady linear pyrolysis"), the kinetic constants can be determined from the values of T_S and the ignition delay, t_{ign}. Some results from the theoretical analysis of this process are presented in [6], while methods for its experimental study are discussed in Chap. 5. In this book, only steady linear pyrolysis is considered (the word "steady" will be omitted from this point on for the sake of brevity).

If the linear pyrolysis rates are lower than or equal to a limiting value (determined by parameters such as the force pressing the sample onto the heater, the sample's cross-sectional area, and the ambient gas pressure), T_S can be considered to be equal to T_0 without significant error. At higher rates of linear pyrolysis, T_S can be estimated provided that the rate of heat flux maintaining pyrolysis is measured along with the heater temperature and the linear pyrolysis rate.

For most practically important compounds decomposing due to the volumetric mechanism [5, 7], the theoretical analysis of linear pyrolysis should result in an analytical expression connecting the linear pyrolysis rate, U, the temperature, T_S, and the kinetic constants.

Fast linear pyrolysis of condensed material characterized by low values of thermophysical parameters under poor outward heat exchange can be described using a model of the linear decomposition of a semi-infinite continuum. In this case, the process takes place in the inner diffusion region [8]. The effective activation energy is equal to the half of the actual activation energy [7]. Experimental data for a zero-order reaction can be analyzed using the Merzhanov formula [9]:

$$U = \sqrt{\frac{ak_0\mathrm{R}T_S^2\exp(-E/\mathrm{R}T_S)}{E\left[(T_S - T_\infty) \mp Q/2c\right]}}, \tag{1.1}$$

where a and c are the thermal diffusivity and heat capacity, k_0 is the preexponential factor in the analytic expression of the Arrhenius law, R is the universal gas constant, E is the activation energy, T_∞ is the ambient temperature, which is identical to the temperature of the opposite sample surface, and Q is the thermal effect of the reaction.

At a low T_S, the linear pyrolysis of many compounds has been found experimentally to occur quite slowly ($\leq 10^{-8}\,\mathrm{m\,s^{-1}}$) without sample distortion. In this case, the dependence of the linear decomposition rate, U, on T_S is determined not only by the kinetic constants of the decomposition reaction and thermophysical parameters of the compound, but also by heat exchange parameters. Under these conditions, the supplied heat is dissipated into the environment due to heat emission from the sample's lateral surface. The difference between the heat supply and heat dissipation is determined by the degree of sample through-heating and by the heat released (absorbed) due to decomposition reactions in the preheat zone.

Let us consider linear decomposition under significant heat losses (Fig. 1.1) and determine a criterion for transition from this regime to a region characterized by negligible outward heat dissipation (the semi-infinite continuum). Barzykin and Khudyaev [10] analyzed the influence of heat emission from the sample's lateral surface upon unsteady linear pyrolysis (critical conditions for ignition).

An exact analytical solution to the nonlinear partial differential equations that describe heat transfer in a two-dimensional region in the presence of a distributed Arrhenius heat source (semi-infinite plate and semi-infinite cylinder) cannot be obtained, even in the simplest case (for a zero-order reaction). Therefore only a

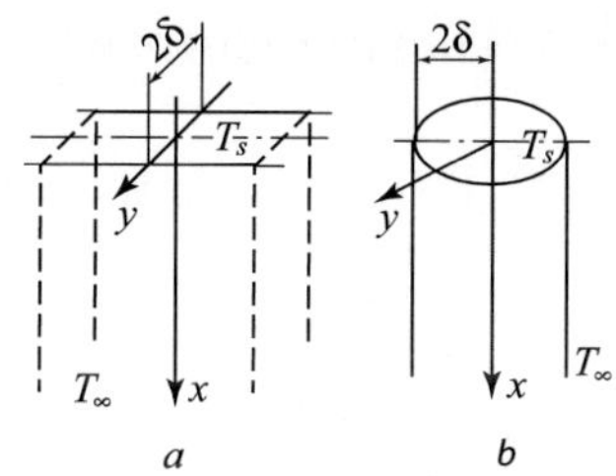

Fig. 1.1 To the problem solution (see the text for the explanations: *a* – for a semi-infinite plate, *b* – for a semi-infinite cylinder

situation where the temperature gradient in the transverse direction is much lower than that in the longitudinal direction is considered ($T'_y << T'_x$ in the sample; in other words, when the thermal resistance is concentrated outside the sample near its lateral surface, and the Bio number, which is characteristic of the heat exchange, $\mathrm{Bi} = \alpha\delta/\lambda << 1$, where α is the coefficient of outward heat exchange, δ is the characteristic size of the sample, and λ is the thermal conductivity coefficient of the sample).

At $\mathrm{Bi} > 1$, the inaccuracy caused by replacing $T = T(x,y)$ with $T = T(x)$ can be eliminated by introducing an averaged temperature, $\overline{T}$ (according to the Frank–Kamenetsky method [8]).

In this case, T should be replaced with $\overline{T}$ in the resulting solution presented below, which does not change the qualitative character of the obtained relationships.

For a one-dimensional problem, the basic equation is

$$\frac{d^2T}{dx^2} + \frac{\mathrm{M}}{\delta}\frac{dT}{dx} \pm \frac{Qk_0}{ac}\exp\left(-\frac{E}{\mathrm{R}T}\right) - \frac{\mathrm{Bi}}{\delta^2}(T - T_\infty) = 0 \tag{1.2}$$

with the boundary conditions

$$T|_{x=0} = T_S, \tag{1.3}$$

$$T|_{x\to\infty} = T_\infty, \tag{1.4}$$

where δ is the half-thickness of the plate (or half-radius of the cylinder), the Michelson number $\mathrm{M} = U\delta/a$, and Bi is the Bio number.

Let us consider the following cases:

- Thermoneutral reaction $Q = 0$
- Reaction with low thermal effect

$$\left|\frac{Q}{c(T_S - T_\infty)}\right| << 1 \tag{1.5}$$

- Reaction with high thermal effect

$$\left|\frac{Q}{c(T_S - T_\infty)}\right| >> 1 \tag{1.6}$$

1.2 Thermoneutral Reaction

In this case, the thermal conductivity equation is

$$\frac{d^2T}{dx^2} + \frac{\mathrm{M}}{\delta}\frac{dT}{dx} - \frac{\mathrm{Bi}}{\delta^2}(T - T_\infty) = 0 \tag{1.7}$$

with the boundary conditions (1.3) and (1.4).

The solution for this equation is

$$T = T_\infty + (T_S - T_\infty)\exp\left[-\frac{\mathrm{M}}{2}\left(1+\sqrt{1+\frac{4\mathrm{Bi}}{\mathrm{M}^2}}\right)\frac{x}{\delta}\right]. \tag{1.8}$$

The temperature distribution in the sample described by Eq. (1.8) is similar to the Michelson temperature distribution [11]

$$T = T_\infty + (T_S - T_\infty)\exp\left(-\frac{xU}{a}\right).$$

The difference in the exponents is associated with the effect of heat emission from the sample's lateral surface. The temperature determined by Eq. (1.8) is denoted as T_M below.

In the case of a zero-order reaction, the rate of linear pyrolysis can be estimated by using the following equation [9]:

$$U = k_0 \int_0^\infty \exp\left[-\frac{E}{\mathrm{R}T(U,x)}\right]dx. \tag{1.9}$$

To calculate the integral, one can use the expansion of the exponential term according to the Frank–Kamenetsky method [8] and the power series expansion of $T = T(x)$. This simplification, which is valid for small x, will be applied to solve this problem using other methods. Far from the limit $x = 0$, the kinetic function $k_0 \exp(-E/\mathrm{R}T)$ is close to zero. Thus, these simplifications do not cause significant errors in the final expression for the integral Eq. (1.9).

Then, from Eq. (1.9),

$$U = \frac{2k_0\delta \mathrm{R}T_S^2 \exp(-E/\mathrm{R}T_S)}{E(T_S - T_\infty)\mathrm{M}\left(1+\sqrt{1+4\mathrm{Bi}/\mathrm{M}^2}\right)}$$

and

$$\mathrm{M} = \sqrt{\frac{2k_0\delta^2 \mathrm{R}T_S^2 \exp(E/\mathrm{R}T_S)}{aE(T_S - T_\infty)\left(1+\sqrt{1+4\mathrm{Bi}/\mathrm{M}^2}\right)}}. \tag{1.10}$$

1.3 Reaction with Low Thermal Effect

In this case, considering Eq. (1.2) in combination with inequality Eq. (1.5), one can assume that for the entire region

$$\frac{\mathrm{M}}{\delta}\frac{dT}{dx} >> \frac{Qk_0}{ac}\exp\left(-\frac{E}{\mathrm{R}T}\right).$$

Now, to solve Eq. (1.2), one can use the perturbation method on the assumption that the temperature profile and the pyrolysis rate for a reaction with low thermal effect are quite similar to those of a thermoneutral reaction (Fig. 1.2). In other words,

$$T(x) = T_M(x) + \Delta T(x) \tag{1.11}$$

and

$$\mathrm{M} = \mathrm{M}_M + \Delta\mathrm{M}, \tag{1.12}$$

where M_M is determined by Eq. (1.10).

When Eqs. (1.11) and (1.12) are substituted into Eqs. (1.2) and (1.7) is subtracted, the result is

$$\frac{d^2\Delta T}{dx^2} + \frac{\mathrm{M}}{\delta}\frac{d\Delta T}{dx} + \frac{\Delta\mathrm{M}}{\delta}\frac{dT_M}{dx} \pm \frac{Qk_0}{ac}\exp\left[-\frac{E}{\mathrm{R}(T_M+\Delta T)}\right] - \frac{\mathrm{Bi}}{\delta^2}\Delta T = 0.$$

This can be rearranged using the approach applied in the case of a thermoneutral reaction. Taking into account that

$$\Delta T << \frac{\mathrm{R}T_S^2}{E}, \tag{1.13}$$

one can obtain the following equation with respect to ΔT:

$$\begin{aligned} &\frac{d^2\Delta T}{dx^2} + \frac{\mathrm{M}}{\delta}\frac{d\Delta T}{dx} + \frac{\Delta\mathrm{M}}{\delta}\frac{dT_M}{dx} \\ &\quad \pm \frac{Qk_0}{ac}\exp\left(-\frac{E}{\mathrm{R}T_S}\right)\exp\left[-\frac{E(T_S-T_\infty)}{\mathrm{R}T_S^2}\mathrm{M}_M\frac{1+\sqrt{1+4\mathrm{Bi}/\mathrm{M}_M^2}}{2}\frac{x}{\delta}\right] \\ &\quad \times \exp\frac{E\Delta T}{\mathrm{R}T_S^2} - \frac{\mathrm{Bi}}{\delta^2}\Delta T = 0 \end{aligned} \tag{1.14}$$

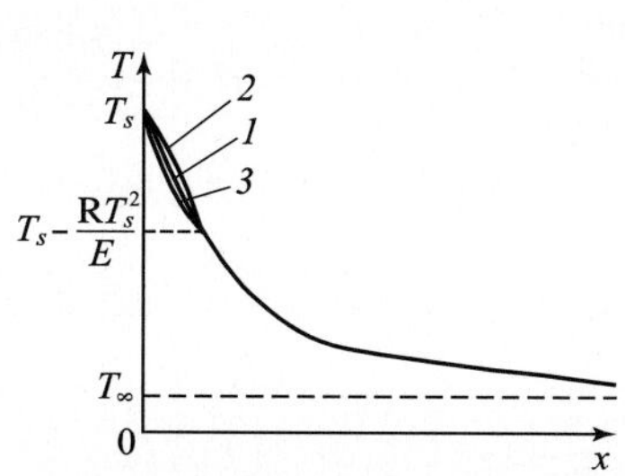

Fig. 1.2 Temperature profiles in the sample for a thermoneutral reaction (*1*), an exothermic reaction (*2*), an endothermic reaction (*3*)

with boundary conditions

$$\Delta T|_{x=0} = 0, \tag{1.15}$$

$$\Delta T|_{x\to\infty} = 0. \tag{1.16}$$

Equation (1.14) can be transformed into a linear one by using inequality Eq. (1.13) as follows:

$$\frac{d^2\Delta T}{dx^2} + \frac{\mathrm{M}}{\delta}\frac{d\Delta T}{dx} - \frac{\mathrm{Bi}}{\delta^2}\Delta T = \frac{\mathrm{M_M}\Delta\mathrm{M}}{\delta^2}(T_S - T_\infty)$$
$$\times \frac{1+\sqrt{1+4\mathrm{Bi}/\mathrm{M}_M^2}}{2}\exp\left(-\mathrm{M}_M\frac{1+\sqrt{1+4\mathrm{Bi}/\mathrm{M}_M^2}}{2}\frac{x}{\delta}\right)$$
$$\pm\frac{Qk_0}{ac}\exp\left(\frac{E}{\mathrm{R}T_S}\right)\exp\left[-\frac{E(T_S-T_\infty)}{\mathrm{R}T_S^2}\mathrm{M_M}\frac{1+\sqrt{1+4\mathrm{Bi}/\mathrm{M}_M^2}}{2}\frac{x}{\delta}\right] \tag{1.17}$$

with boundary conditions Eqs. (1.15) and (1.16). After some rearrangement, the solution to Eq. (1.17) is

$$\Delta T = (T_S - T_\infty)\left[\exp\left(-\mathrm{M}\frac{1+\sqrt{1+4\mathrm{Bi}/\mathrm{M}^2}}{2}\frac{x}{\delta}\right)\right.$$
$$\left.-\exp\left(-\mathrm{M_M}\frac{1+\sqrt{1+4\mathrm{Bi}/\mathrm{M}_M^2}}{2}\frac{x}{\delta}\right)\right]$$
$$\pm\frac{Qk_0\exp(-E/\mathrm{R}T_S)}{ac\,[\beta^2-(\mathrm{M}/\delta)\beta-\mathrm{Bi}/\delta^2]}$$
$$\times\left[\exp\left(-\mathrm{M}\frac{1+\sqrt{1+4\mathrm{Bi}/\mathrm{M}^2}}{2}\frac{x}{\delta}\right)-\exp(-\beta x)\right], \tag{1.18}$$

where

$$\beta = -\frac{E(T_S-T_\infty)}{\mathrm{R}T_S^2}\frac{\mathrm{M}_M}{\delta}\frac{1+\sqrt{1+4\mathrm{Bi}/\mathrm{M}_M^2}}{2}.$$

It can be shown that the following relationships are always valid:

$$\frac{\mathrm{M}}{\delta\beta} << 1 \tag{1.19}$$

and

$$\frac{\mathrm{Bi}}{\delta^2\beta^2} << 1. \tag{1.20}$$

Combining Eqs. (1.8), (1.11), and (1.18)–(1.20), one can obtain for the temperature profile in the sample

$$T - T_\infty = (T_S - T_\infty)\exp\left(-\mathrm{M}\frac{1+\sqrt{1+4\mathrm{Bi}/\mathrm{M}^2}}{2}\frac{x}{\delta}\right)$$
$$\pm\frac{Qk_0}{ac\beta^2}\exp\left(-\frac{E}{\mathrm{R}T_S}\right)\left[\exp\left(-\mathrm{M}\frac{1+\sqrt{1+4\mathrm{Bi}/\mathrm{M}^2}}{2}\frac{x}{\delta}\right) - \exp(-\beta x'')\right].$$

Then one can obtain an approximate expression for $(T_S - T)$:

$$T_S - T = (T_S - T_\infty)\mathrm{M}\frac{1+\sqrt{1+4\mathrm{Bi}/\mathrm{M}^2}}{2}\frac{x}{\delta}$$
$$\mp\frac{2Qk_0\mathrm{R}T_S^2\delta^2\exp(-E/\mathrm{R}T_S)}{acE(T_S - T_\infty)\mathrm{M}_M(1+\sqrt{1+4\mathrm{Bi}/\mathrm{M}^2})}\frac{x}{\delta}$$
$$= (T_S - T_\infty)\mathrm{M}\frac{1+\sqrt{1+4\mathrm{Bi}/\mathrm{M}^2}}{2}\frac{x}{\delta}\mp\frac{Q\mathrm{M}_\mathrm{M}}{\mathrm{c}}\frac{x}{\delta}. \tag{1.21}$$

Now one can write an expression for the pyrolysis rate by using Eq. (1.9) and assuming $\mathrm{M}_M \sim \mathrm{M}$ for the second term on the right hand side of Eq. (1.21):

$$U = \frac{k_0\delta\mathrm{R}T_S^2\exp(-E/\mathrm{R}T_S)}{\mathrm{M}E\left[(T_S - T_\infty)\left(1+\sqrt{1+4\mathrm{Bi}/\mathrm{M}^2}\right)/2\mp Q/c\right]},$$

and then one finds that

$$\mathrm{M} = \sqrt{\frac{k_0\delta^2\mathrm{R}T_S^2\exp(-E/\mathrm{R}T_S)}{aE\left[(T_S - T_\infty)\left(1+\sqrt{1+4\mathrm{Bi}/\mathrm{M}^2}\right)/2\mp Q/c\right]}}. \tag{1.22}$$

1.4 Reaction with High Thermal Effect

This case can be considered in terms of two zones (Fig. 1.3): a reaction zone I and a preheat zone II. For zone I, the thermal conductivity equation term related to the first temperature derivative is usually neglected [8, 12]. In this case, this term can

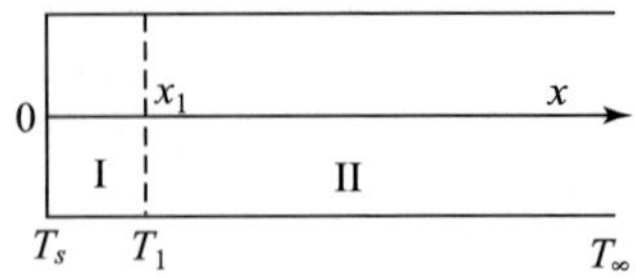

Fig. 1.3 Sample zones: I, reaction occurs with a noticeable rate; II, preheat zone

be taken into account under the assumption that it is approximately equal to that of the equation for a thermoneutral reaction. Then one can write a thermal conductivity equation as

$$\frac{d^2T}{dx^2}+\frac{\mathrm{M}}{\delta}\frac{dT}{dx}\pm\frac{Qk_0}{ac}\exp\left(-\frac{E}{\mathrm{R}T}\right)-\frac{\mathrm{Bi}}{\delta^2}(T-T_\infty)=0. \tag{1.23}$$

Integration of this equation over the interval 0 to x_1 gives

$$\left(\frac{dT}{dx}\right)^2\Bigg|_{x=x_1}=\frac{q^2}{\lambda^2}-\left(\frac{\mathrm{Bi}}{\delta^2}+\frac{\mathrm{M}^2}{\delta^2}\frac{1+\sqrt{1+4\mathrm{Bi}/\mathrm{M}^2}}{2}\right)$$
$$\times\left[(T_S-T_\infty)^2-(T_1-T_\infty)^2\right]\pm\frac{2Qk_0\mathrm{R}T_S^2}{acE}\exp\left(-\frac{E}{\mathrm{R}T_S}\right), \tag{1.24}$$

where x_1 is the thickness of the reaction zone, q is the heat flow towards the end surface, and T_1 is the temperature on the border between the two zones.

To find q, one can use the heat-balance equation:

$$q=U\rho\left[c(T_S-T_\infty)\mp Q\right]+\frac{\alpha}{\delta}\int_0^\infty(T-T_\infty)dx.$$

The contribution of the heat emitted from the lateral sample surface in the reaction zone to the overall heat loss is always much less than that of the preheat zone. Therefore, during the estimation of the overall heat loss, one can substitute T for T_M in the integral without causing significant error. Then

$$\int_0^\infty(T-T_\infty)dx=(T_S-T_\infty)\mathrm{M}\delta\frac{\sqrt{1+4\mathrm{Bi}/\mathrm{M}^2}-1}{2\mathrm{Bi}}, \tag{1.25}$$

and

$$\frac{q}{\lambda}=\frac{\mathrm{M}}{\delta}\left[(T_S-T_\infty)\frac{1+\sqrt{1+4\mathrm{Bi}/\mathrm{M}^2}}{2}\mp\frac{Q}{c}\right]. \tag{1.26}$$

Let us consider the first two terms on the right hand side of Eq. (1.24):

$$\frac{q}{\lambda^2} - \left(\frac{\text{Bi}}{\delta^2} + \frac{\text{M}^2}{\delta^2}\frac{1+\sqrt{1+4\text{Bi}/\text{M}^2}}{2}\right)\left[(T_S - T_\infty)^2 - (T_1 - T_\infty)^2\right]$$

$$= \frac{\text{M}^2}{\delta^2}(T_1 - T_\infty)^2\left(\frac{1+\sqrt{1+4\text{Bi}/\text{M}^2}}{2}\right)^2$$

$$\mp\frac{2\text{M}^2}{\delta^2}\frac{Q}{c}\left[(T_S - T_\infty)\frac{1+\sqrt{1+4\text{Bi}/\text{M}^2}}{2} \mp \frac{Q}{2c}\right]. \tag{1.27}$$

Substitution of Eq. (1.27) into Eq. (1.24) gives

$$\left.\left(\frac{dT}{dx}\right)^2\right|_{x=x_1} = \frac{\text{M}^2}{\delta^2}(T_1 - T_\infty)^2\left(\frac{1+\sqrt{1+4\text{Bi}/\text{M}^2}}{2}\right)^2 \mp \frac{2\text{M}^2 Q}{\delta^2 c}$$
$$\times\left[(T_S - T_\infty)\frac{1+\sqrt{1+4\text{Bi}/\text{M}^2}}{2} \mp \frac{Q}{2c}\right] \tag{1.28}$$
$$\pm\frac{2Qk_0\text{R}T_S^2}{acE}\exp\left(-\frac{E}{\text{R}T_S}\right).$$

Equation (1.7) with boundary conditions Eq. (1.4) and $T|_{x=x_1} = T_1$ is the energy equation for zone II. The solution of this equation is

$$T = T_\infty + (T_1 - T_\infty)\exp\left(-\text{M}\frac{1+\sqrt{1+4\text{Bi}/\text{M}^2}}{2}\frac{x-x_1}{\delta}\right), \tag{1.29}$$

from which

$$\left.\frac{dT}{dx}\right|_{x=x_1} = -\frac{\text{M}}{\delta}\frac{1+\sqrt{1+4\text{Bi}/\text{M}^2}}{2}(T_1 - T_\infty).$$

If the temperature gradients on the border between the two zones are equal,

$$\frac{\text{M}^2}{\delta^2}(T_1 - T_\infty)^2\left(\frac{1+\sqrt{1+4\text{Bi}/\text{M}^2}}{2}\right)^2 = \frac{\text{M}^2}{\delta^2}(T_1 - T_\infty)^2\left(\frac{1+\sqrt{1+4\text{Bi}/\text{M}^2}}{2}\right)^2$$
$$\mp\frac{2\text{M}^2}{\delta^2}\frac{Q}{c}\left[(T_S - T_\infty)\frac{1+\sqrt{1+4\text{Bi}/\text{M}^2}}{2} \mp \frac{Q}{2c}\right] \pm \frac{2Qk_0\text{R}T_S^2}{acE}\exp\left(-\frac{E}{\text{R}T_S}\right). \tag{1.30}$$

This gives

$$M = \left(\frac{k_0 R T_S^2 \delta^2 \exp(-E/RT_S)}{aE\left[(T_S - T_\infty)\left(1+\sqrt{1+4\mathrm{Bi}/M^2}\right)/2 \mp Q/2c\right]} \right)^{1/2}. \tag{1.31}$$

This is the most general expression. Relationship Eq. (1.22), when obtained using a perturbation method, gives an identical dependence for the linear decomposition rate. The difference in the coefficients before term Q/c is explained by the double series expansion of the exponential term.

The resulting Eq. (1.31) can be obtained in a shorter (though less rigorous) way if the heat losses due to the warming-up of the sample and heat emission from the lateral surface of the sample in the first zone are neglected. In this case, the thermal conductivity equation (Eq. 1.23) for the first zone is

$$\frac{d^2T}{dx^2} \pm \frac{Qk_0}{ac} \exp\left(-\frac{E}{RT}\right) = 0,$$

while for the second zone it does not change. After mathematical manipulations similar to those performed for Eqs. (1.24)–(1.30), and taking into account the equality of the heat flows at the boundary between the two zones, one obtains

$$\left[\frac{M}{\delta}\frac{1+\sqrt{1+4\mathrm{Bi}/M^2}}{2}(T_S - T_\infty)\right]^2 = \frac{M^2}{\delta^2}\left[(T_S - T_\infty)\frac{1+\sqrt{1+4\mathrm{Bi}/M^2}}{2} \mp \frac{Q}{c}\right]^2 \pm \frac{2Qk_0RT_S^2}{acE}\exp\left(-\frac{E}{RT_S}\right).$$

Equation (1.31) also represents the solution to this equation.

1.5 Analysis of the Basic Expression for the Linear Pyrolysis Rate

Solution of Eq. (1.31) with respect to M results in a very bulky expression. It is more convenient to analyze the physical meaning of relationship (1.31) for two limiting cases:

1. High rates of linear decomposition (burning) and insignificant heat losses from the lateral surface; that is

$$\frac{4\mathrm{Bi}}{M^2} << 1, \tag{1.32}$$

2. Low rates of linear decomposition and significant heat losses; that is

$$\frac{4\mathrm{Bi}}{\mathrm{M}^2} >> 1. \tag{1.33}$$

For the first case, one can obtain the Merzhanov formula (Eq. 1.1) under the assumption that Bi = 0, that no heat emission from the lateral surface occurs (for a first approximation). The rate determined by using this expression is denoted $U*$, and the corresponding Michelson number

$$\mathrm{M}^* = \frac{U^*\delta}{a}. \tag{1.34}$$

Then, for the second approximation one can obtain

$$\mathrm{M} = \mathrm{M}^*\sqrt{1 - \frac{\mathrm{Bi}}{\mathrm{M}^{*2}}\frac{1}{1 \mp Q/[2c - (T_S - T_\infty)]}}. \tag{1.35}$$

For a thermoneutral reaction this relationship can be written as

$$\mathrm{M} = \mathrm{M}^*\sqrt{1 - \frac{\mathrm{Bi}}{\mathrm{M}^{*2}}}. \tag{1.36}$$

Analysis of Eqs. (1.1), (1.35), and (1.36) leads to the conclusion that, at high rates of linear decomposition accompanied by insignificant outward heat losses, the dependence of the process rate on the surface temperature, the kinetic constants of the reaction and the thermophysical characteristics of the material is similar to that for semi-infinite continuum decomposition; that is

$$U \sim \exp\left(-\frac{E}{2RT_S}\right).$$

The expression obtained for the second limiting case is especially interesting, since (as was mentioned earlier) the decomposition of a wide range of systems can occur quite slowly.

Taking into consideration condition Eq. (1.33), one can obtain the following relationship from the basic Eq. (1.31) in the first approximation:

$$\mathrm{M} = \frac{\delta^2 k_0 \mathrm{R} T_S^2 \exp(-E/\mathrm{R}T_S)}{aE(T_S - T_\infty)\sqrt{\mathrm{Bi}}}$$

or, by using definition Eq. (1.34),

$$\mathrm{M} = \frac{\mathrm{M}}{\sqrt{\mathrm{Bi}}}\left[1 \mp \frac{Q}{2c(T_S - T_\infty)}\right].$$

For a thermoneutral reaction this expression can be written as

$$\mathrm{M} = \frac{\mathrm{M}^{*2}}{\sqrt{\mathrm{Bi}}}.$$

Thus, for the dimensional pyrolysis rate one obtains

$$U = \frac{\delta \mathrm{R} T_S^2 k_0 \exp(-E/\mathrm{R}T_S)}{E(T_S - T_\infty)\sqrt{\mathrm{Bi}}}. \tag{1.37}$$

In this case, the dependence of the decomposition rate on the surface temperature is qualitatively different:

$$U \sim \exp\left(-\frac{E}{\mathrm{R}T_S}\right).$$

This is connected to the difference in the dependences of the reaction zone width on the process parameters. For the first case, the expression for the reaction zone width, Δ_I (accurate up to $\frac{\mathrm{R}T_S^2}{E(T_S - T_\infty)}$, which can be assumed to be constant), is

$$\Delta_\mathrm{I} \sim \frac{a}{U}. \tag{1.38}$$

In other words, Δ_I unrestrictedly increases as the process rate (the temperature) decreases. In contrast, in the second case, the width of the reaction zone (accurate up to the same factor) is

$$\Delta_\mathrm{II} \sim \frac{\delta}{\sqrt{\mathrm{Bi}}}, \tag{1.39}$$

and is almost constant, since it depends only on the heat loss-determining size of the sample, δ, and the Bi number (the parameter determining the outward heat exchange). Parameter $\frac{\delta}{\sqrt{\mathrm{Bi}}}$ is the characteristic depth of the preheat zone for a semi-infinite rod heated from the end [13]. The quantity corresponds to a decrease in the dimensionless temperature by factor of e.

The comments presented above allow one to understand the physical meaning of expression Bi/M^2.

From expressions (1.38) and (1.39), one can see that

$$\frac{\sqrt{\mathrm{Bi}}}{\mathrm{M}} = \frac{\Delta_\mathrm{I}}{\Delta_\mathrm{II}}.$$

Under condition (1.32), the thickness of the reaction layer is significantly smaller than the characteristic thickness of the preheat zone. As the pyrolysis rate decreases (within the limits of condition 1.32), the thickness of the reaction layer increases, heat emission from the lateral surface is not significant, and the dependence of the process rate on the surface temperature is almost identical to that of the process occurring without heat losses from the lateral surface.

Under condition (1.33), the thickness of the reaction layer reaches its maximum value and remains constant upon decreasing the sample surface temperature (and the decomposition rate) due to its connection to the constant thickness of the preheat zone.

Thus, as the surface temperature and the rate of linear decomposition decrease, the process of pyrolysis moves from the inner diffusion region into the inner kinetic region, which is characterized by identical macroscopic and actual kinetics for the decomposition reaction.

1.5.1 Linear Pyrolysis of Thin ("Finite Length") Samples

Having solved Eq. (1.2) with boundary condition (1.3) and additional condition $T|_{x=l} = T_\infty$, one can show that expression (1.1) can be used for the analysis of limiting case (1.32) if $l >> \frac{a}{U}$, while in the case of Eq. (1.33), Eq. (1.37) can be used if $l >> \frac{\delta}{\sqrt{\mathrm{Bi}}}$. In the case of the linear pyrolysis of a very thin sample, the expression for the dimensional rate is

$$U = \frac{l\mathrm{R}T_S^2 k_0 \exp(-E/\mathrm{R}T_S)}{E(T_S - T_\infty)}. \quad (1.40)$$

This relationship is of significant interest for the practical application of linear pyrolysis for studying process kinetics, since it does not contain the thermophysical characteristics of a decomposing compound (it is very difficult to determine these characteristics for the high-temperature region).

Figure 1.4 schematically illustrates an experimental approach that can be used to study the linear pyrolysis of a thin sample (T_S and T_∞ are the temperatures maintained at the ends of the sample). Based on the similarity of the triangles, the thickness of the reaction layer, Δ, which corresponds to the temperature change of one Semenov interval, $\mathrm{R}T_S^2/E$, is related to the diminishing thickness of the sample, l, as

$$\Delta = l\frac{\mathrm{R}T_S^2}{E(T_S - T_\infty)}. \quad (1.41)$$

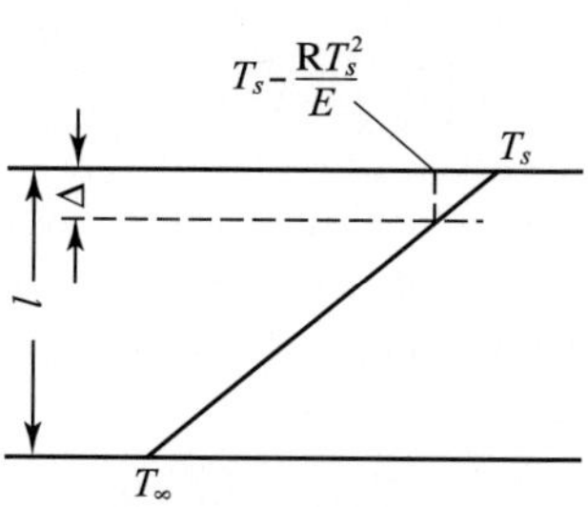

Fig. 1.4 Derivation of expression (1.40)

Since the rate of linear steady propagation of the reaction front always equals the product of the reaction layer thickness and the rate constant of the volumetric reaction at the characteristic temperature, one obtains expression (1.40).

Noting that, by definition $U = -\mathrm{d}l/\mathrm{d}t$ for linear pyrolysis under quasi-steady conditions, one can integrate relationship (1.40) and obtain the expression for the time dependence of the sample thickness:

$$l = l_0 \exp\left(-\frac{t}{t_0}\right), \tag{1.42}$$

where l_0 is the initial thickness of the sample and

$$t_0 = \frac{E(T_S - T_\infty)}{RT_S^2 k_0} \exp\frac{E}{RT_S}. \tag{1.43}$$

Thus, despite the complicated nature of the experimental technique, the use of the linear pyrolysis-based experimental approach in combination with the analysis of the results using Eqs. (1.40)–(1.42) allows one to obtain data on the kinetics of the volumetric reaction without the need for data on the thermophysical characteristics of the material in the high-temperature region. These data are almost always unavailable. In addition, the limiting stage of the decomposition reaction (if the process occurs in the volume or on the surface) can be directly identified. In the latter case (evaporation, dissociative sublimation, etc.), the rate of linear pyrolysis does not decrease with decreasing sample thickness.

1.6 Effect of Conversion Degree on the Linear Pyrolysis Rate

In order to take burn-out during linear pyrolysis into account, one may write

$$\frac{d^2T}{dx^2} + \frac{\mathrm{M}}{\delta}\frac{dT}{dx} \pm \frac{Qk_0\varphi(\eta)}{ac}\exp\left(-\frac{E}{RT}\right) - \frac{\mathrm{Bi}}{\delta^2}(T - T_\infty) = 0, \tag{1.44}$$

$$U\frac{d\eta}{dx} + k_0\varphi(\eta)\exp\left(-\frac{E}{RT}\right) = 0 \tag{1.45}$$

with the boundary conditions (1.3)–(1.4) and

$$\eta|_{x=0} = 1 - \eta_g, \tag{1.46}$$

$$\eta|_{x\to\infty} = 0. \tag{1.47}$$

For the linear pyrolysis of a "finite" length sample, the boundary conditions for the cold end ($x = l$) are

$$T|_{x=l} = T_\infty,\ \eta|_{x=l} = 0.$$

Here η_g is the part of the unreacted compound that leaves the pyrolysis surface due to the physical process accompanying decomposition (for example dispersion, which includes processes such as dissociative sublimation and evaporation that result in the partial loss of unreacted compound from the sample surface, or sublimation).

The set of Eqs. (1.44)–(1.47) can be solved using methods similar to those applied to analyze linear pyrolysis in the case of zero-order reactions (Sects. 1.1–1.5) under the standard assumptions made for the analysis of condensed system burning accompanied by dispersion or sublimation [14, 15, 16, 17, 18].

After a number of mathematical manipulations (leading to bulky expressions omitted for the sake of brevity), one can write final relationships for M corresponding to two processes:

linear pyrolysis in the case of the zero-order reaction $[\varphi(\eta) = 1]$ accompanied by dispersion,

$$\mathrm{M} = \left\{ \frac{\delta^2 \mathrm{R} T_S^2 k_0 \exp(-E/\mathrm{R}T_S)}{aE\left[(T_S - T_\infty)\dfrac{1+\sqrt{1+4\mathrm{Bi}/\mathrm{M}^2}}{2}(1-\eta_g) \mp \dfrac{Q}{2c}(1-\eta_g)^2\right]} \right\}^{1/2} \tag{1.48}$$

and linear pyrolysis in the case of the zero-order reaction $[\varphi(\eta) = 1 - \eta]$,

$$\mathrm{M} = \left\{ \frac{\delta^2 \mathrm{R} T_S^2 k_0 \exp(-E/\mathrm{R}T_S)}{aE\left[(T_\infty - T_S)\dfrac{1+\sqrt{1+4\mathrm{Bi}/\mathrm{M}^2}}{2}\ln\eta_g \pm \dfrac{Q}{2c}(1-\eta_g+\ln\eta_g)^2\right]} \right\}^{1/2}. \tag{1.49}$$

Expressions for the linear pyrolysis rates for the two modes that differ in terms of the conditions of outward heat exchange can easily be obtained from relationships (1.48) and (1.49) (Table 1.1).

Joint analysis of the expressions for the linear pyrolysis rates (Table 1.1) allows one to evaluate the dispersion contribution into an overall rate of linear decomposition for various volumetric reactions.

At $\eta_g \approx 1$ Eqs. (50), (51) as well as (52), (53) and (54), (55) are identical; in other words, at significant dispersion (or at a very low burn-out in the surface layer) the linear pyrolysis rate does not depend on the reaction order. At insignificant dispersion ($\eta_g \ll 1$) the effect of the unreacted compound loss on the linear pyrolysis rate can be neglected for a zero-order reaction. In the case of first-order reaction (in general at $n \neq 0$, where n is the reaction order), the burn-out markedly influences the linear pyrolysis rate. Plots of the dimensionless rate of linear pyrolysis for zero-order

Table 1.1 Analytical expressions for the rate of linear pyrolysis in the limiting modes

Mode of linear pyrolysis	For zero-order reaction		For first-order reaction	
"Fast" linear pyrolysis $\left(\frac{4\mathrm{Bi}}{\mathrm{M}^2} << 1\right)$	$U = \left\{ \frac{ak_0\mathrm{R}T_S^2 \exp\left(-\frac{E}{\mathrm{R}T_S}\right)}{E\left[(T_S - T_\infty)(1-\eta_g) \mp \frac{Q}{2c}(1-\eta_g)^2\right]} \right\}^{1/2}$	(1.50)	$U = \left\{ \frac{ak_0\mathrm{R}T_S^2 \exp\left(-\frac{E}{\mathrm{R}T_S}\right)}{E\left[(T_\infty - T_S)\ln\eta_g \pm \frac{Q}{2c}(1-\eta_g+\ln\eta_g)\right]} \right\}^{1/2}$	(1.51)
"Slow" linear pyrolysis for:				
a semi-infinite sample $\left(\frac{4\mathrm{Bi}}{\mathrm{M}^2} >> 1\right)$	$U = \frac{\delta k_0\mathrm{R}T_S^2 \exp(-E/\mathrm{R}T_S)}{E(T_S - T_\infty)(1-\eta_g)\sqrt{\mathrm{Bi}}}$	(1.52)	$U = \frac{\delta k_0\mathrm{R}T_S^2 \exp(-E/\mathrm{R}T_S)}{E(T_S - T_\infty)\ln\eta_g\sqrt{\mathrm{Bi}}}$	(1.53)
a sample of a "finite" length $\left(l << \frac{U}{a}\right)$	$U = \frac{lk_0\mathrm{R}T_S^2 \exp(-E/\mathrm{R}T_S)}{E(T_S - T_\infty)(1-\eta_g)}$	(1.54)	$U = \frac{lk_0\mathrm{R}T_S^2 \exp(-E/\mathrm{R}T_S)}{E(T_\infty - T_S)\ln\eta_g}$	(1.55)

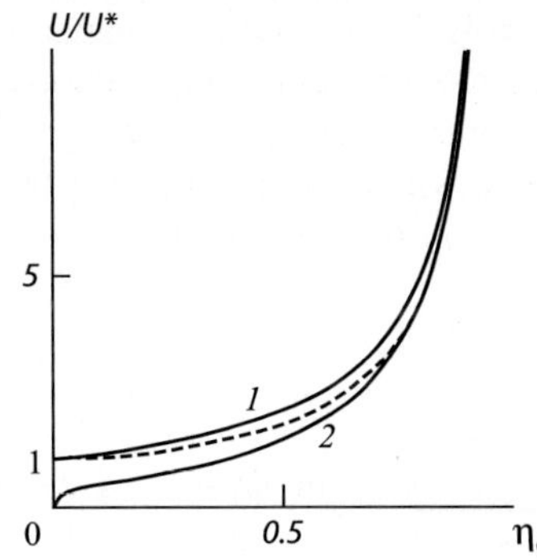

Fig. 1.5 Plot of the relative rate of linear pyrolysis against the dispersion degree for a zero-order reaction (*1*) and for a first-order reaction (*2*)

and first-order reactions against the dispersion degree are shown in Fig. 1.5 [$U*$ is expressed by relationships (1.1) and (1.37) for linear pyrolysis occurring without dispersion in the case of zero-order reactions].

To conclude this section, the following general comment should be made: the macrokinetics of decomposition characterized by a simple (not self-accelerating) reaction mechanism and occurring during the "slow" linear pyrolysis of a sample on a smooth solid metal heater can be described in terms of a pseudo-zero-order (or similar) process. This can be explained by the continuous densification of the surface layer of the substance due to the force pressing the sample onto the heater. At high rates of decomposition ("fast" linear pyrolysis) accompanied by significant dispersion of the substance, the pressure between the sample and the heater (in standard experiments, the applied pressure is usually in the range of one to a few atm; see Chap. 3) almost does not influence processes in the surface layer. These factors should be taken into account when estimating kinetic constants from experimental data on linear pyrolysis. Analysis of experimental data in the case of fast linear pyrolysis with significant dispersion and minor densification of the surface layer by using the Arrhenius anamorphosis approach can result in an apparent increase in the kinetic constants k_0 and E. The qualitative dependence of the linear pyrolysis rate on the degree of dispersion corresponding to linear pyrolysis under real conditions is shown in Fig. 1.5 as a dotted line.

1.7 Effect of the Gas-Film on Linear Pyrolysis: Method for Calculating the Surface Temperature

One of the most challenging tasks one has to face when studying the kinetics of the fast decomposition of condensed compounds is to determine the characteristic sample temperature. The latter must be dramatically different from the temperature of the thermostat used. For the kinetic analysis of slow linear pyrolysis ($U \leq 10^{-3}\,\mathrm{mm\,s^{-1}}$), the temperature drop across the heater–sample contact can be neglected assuming that $T_S = T_0$. However, for fast process (for example, the linear pyrolysis of polymeric materials), the temperature of the hot plate, T_0, separated from the decomposing compound by a gas layer can significantly exceed the temperature of the pyrolysis surface, T_S. This effect was first mentioned by Cantrell [19], who analyzed the hydrodynamics of the gas layer and obtained the following relationship between its thickness, Z, and basic parameters of the process:

$$Z = \left(\frac{3\pi}{32} \frac{U \rho_s g \nu_g b^4}{W} \right)^{1/3}, \tag{1.56}$$

where ρ_s is the density of the solid compound, ν_g is the viscosity of the gas produced by linear pyrolysis, b is the sample diameter, and W is the force pressing the sample to the heater. Cantrell showed that the heat transfer to the surface of the reacting compound occurs mostly due to conduction, estimated the T_S for the linear pyrolysis of solid carbonic acid at $T_0 \approx 50°C$, and found out that the temperature drop across the gas gap can be as high as 100°C. For his experimental and theoretical studies, Cantrell used data obtained by direct calorimetric measurements of the heat flow maintaining the linear pyrolysis and highly accurate thermophysical parameters for carbonic acid gas. The results of Cantrell's work were not used in subsequent studies of linear pyrolysis for compounds that are important from a practical viewpoint (specifically polymers). Most likely, this was usually caused by the lack of data on the thermophysical parameters, the viscosity and the composition of the gap-filling gas produced by linear pyrolysis. In addition, calorimetric equipment used for studying high-temperature linear pyrolysis experimentally is much more sophisticated than the experimental setup used by Cantrell. In fact, the application of high-temperature calorimetry eliminates some of the main advantages of the method based on linear pyrolysis: its simplicity and reliability. Therefore, the direct calculation of the pyrolysis surface temperature, T_S, using experimental data on the linear pyrolysis rate, U, at a certain temperature of the heater, T_0, is of great interest. A method of calculating this is given below.

The equation for the balance of heat maintaining linear pyrolysis is

$$\lambda_g \frac{T_0 - T_S}{Z} = \dot{m} \left[c_s (T_S - T_\infty) + Q \right], \tag{1.57}$$

where $\dot{m} = U \rho_s$ is the mass rate of linear pyrolysis, λ is the thermal conductivity of the gas produced by linear pyrolysis, and c_s is the thermal conductivity of the solid compound. Substituting Eq. (1.56) into Eq. (1.57) and neglecting the temperature dependence of the thermophysical parameters of the gas in the first approximation, one can obtain

$$B \frac{(\dot{m} b)^{4/3}}{W^{1/3}} = \left(\frac{32}{3\pi} \right)^{1/3} \frac{\lambda_g (T_0 - T_S)}{\nu^{1/3} \left[c_s (T_S - T_\infty) + Q \right]}, \tag{1.58}$$

where $B = (W/SP_\infty + 1)^{-1/3} = \text{const}$ and S is the area of the linear pyrolysis surface.

The situation of a weak dependence of T_S on increasing T_0 and U at fast linear pyrolysis corresponds to a linear relationship between the left hand side of Eq. (1.58) and T_0. A plot of $W^{-1/3}(\dot{m}b)^{4/3}B$ against T_0 is linear, and the approximate surface temperature can be found from the intercept on the x-axis. The slope ratio, ψ, depends on the gross decomposition heat, the thermal conductivity and the viscosity of the gaseous reaction products. A linear increase in $W^{-1/3}(\dot{m}b)^{4/3}B$ as the heater temperature, T_0, grows indicates a weak increase in the surface temperature, T_S, under these conditions.

Assuming a weak temperature dependence of the thermal-conductivity coefficient of the gas layer, one can determine a set of surface temperature values, T_{Si}, corresponding to a set of heater temperature values, T_{0i}, from the intercepts on the x-axis made by straight lines of slope ψ drawn through the points with coordinates T_{0i} (Fig. 3.11).

1.8 Linear Pyrolysis Due to a Surface Mechanism

The linear pyrolysis of compounds for which gas-producing decomposition occurs in the monomolecular surface layer is of some interest. Indeed, the first experiments on linear pyrolysis were carried out using solid carbonic acid [19] and ammonium haloids [20], which decompose and evolve gas according to a surface mechanism. This aspect is also important when analyzing the contribution of dissociative sublimation to the macrokinetics of the high-temperature linear pyrolysis of ammonium perchlorate (see Chap. 3). The pyrolysis of condensed compounds with the limiting stage of decomposition (dissociation) on the surface is quite similar to equilibrium sublimation. The macrokinetics of pyrolysis are not directly determined by the kinetic parameters of decomposition, while the characteristics of reaction product transport from the compound–heater contact zone are of significant importance. At high rates of linear pyrolysis, the role of the temperature drop across the gas layer becomes more important. In this case, one should use approaches similar to that presented above to determine the real values of T_S. The comparison of the characteristics of linear pyrolysis in the case of an equilibrium surface reaction with those of linear pyrolysis in the case of a volumetric reaction is of great interest. At low rates of linear pyrolysis ($T_S \approx T_\infty$), the removal of the reaction products from the hot zone does not affect the rate of decomposition in volume. The rate of linear pyrolysis increases as the sample diameter increases due to a decrease in the outward heat loss and a corresponding increase in the reaction layer thickness: $U \sim \frac{b}{\sqrt{\mathrm{Bi}}} \sim \sqrt{b}$. In contrast, for systems with a surface equilibrium dissociation reaction, the reaction layer thickness does not depend on the sample size. However, the latter, which is related to the characteristics of the hydrodynamic and diffusion-assisted transport of gaseous reaction products from the reaction zone, influences the rate of the overall process. In other words, the rate of linear pyrolysis decreases as the sample diameter increases. For the linear pyrolysis of a "finite length" sample (see Sect. 1.5), the rate of the process decreases with decreasing l for volumetric reactions, while for reactions with a surface mechanism it should remain constant. Taking these remarks into consideration, one can derive expressions connecting the rate of linear pyrolysis for an equilibrium dissociation reaction with parameters characterizing the thermodynamics of "solid reagent →gaseous products" reactions by making use of methods developed in the theory of transport processes in dispersion systems [21].

In the case of the practically interesting dissociative sublimation of ammonium salts with the formation of ammonia and acid, these expressions are

$$U \approx \frac{Dr\sqrt{A}}{\rho \mathrm{R} T_S^2 b^2} \exp\left(-\frac{\Delta H}{2\mathrm{R}T_S}\right) \tag{1.59}$$

and

$$U \approx \frac{r\sqrt{A}}{\xi b^2} \exp\left(-\frac{\Delta H}{2\mathrm{R}T_S}\right), \tag{1.60}$$

where A is the preexponential factor in the expression for the equilibrium constant, ξ is the coefficient of resistance in the Darcy law, ΔH is the dissociation heat, and r is the radius (effective) of the escape channel for the gas.

Relationship (1.59) corresponds to a relatively low pressure of gaseous products from the thermodynamically equilibrium process, P_S ($P_S \ll P_\infty$). The reaction products are eliminated from the reaction zone due to diffusion-assisted transport (D is the diffusion coefficient). Relationship (1.60) is valid in the case of relatively high P_S ($P_S \geq P_\infty$) for hydrodynamic transport of the reaction products.

References

1. Schultz RD, Dekker AO (1955) In: Proceedings of the 5th Symposium on Combustion. Reinhold, New York
2. Beckstead MW (2000) In: 29th JANNAF Propellant Development & Characterisation Meeting, Cocoa Beach, FL, USA, 8–12 May 2000
3. Miller RS (1996) In: Brill TB, Russel TP, Tao WC, Wardle RB (eds) Decomposition, combustion, and detonation chemistry of energetic materials. Materials Research Society, Pittsburgh, PA, p 3
4. Bakhman NN, Belyaev AF (1967) Combustion of heterogeneous condensed systems. Nauka, Moscow
5. Shteinberg AS (1980) In: Merzhanov AG (ed) Heat and mass exchange in combustion processes. Nauka, Chernogolovka, p 85
6. Averson AE, Barzykin VV, Merzhanov AG (1965) J Eng Phys Thermophys 2:172
7. Shteinberg AS, Sokolova NA (1964) Dokl Phys Chem 158:889
8. Frank-Kamenetsky DA (1969) Diffusion and heat transfer in chemical kinetics. Plenum, New York
9. Merzhanov AG, Khaikin BI (1988) Progr Energy Combust Sci 14:1
10. Barzykin VV, Khudyaev SI (1956) Dokl Phys Chem 169:547
11. Mikhelson VA (1930) Collected works, vol 1. Novyi Agronom, Moscow
12. Zel'dovich JaB, Frank-Kamenetsky DA (1938) Dokl Akad Nauk SSSR 19:693
13. Mikheev MA (1956) Fundamentals of heat transfer. GEI, Moscow
14. Merzhanov AG, Dubovitsky FI (1959) Dokl Phys Chem 129:917
15. Merzhanov AG (1960) Dokl Akad Nauk SSSR 6:1439

16. Manelis GB, Strunin VA (1972) In: Stesik LN (ed) Combustion and explosion. Nauka, Moscow, p 53
17. Manelis GB, Strunin VA (1971) Combust Flame 17:69
18. Strunin VA (1965) Rus J Phys Chem 39:222
19. Cantrell RH (1963) AIAAJ 1:1544
20. Chaiken RF (1962) J Chem Phys 37:2311
21. Chudnovsky AF (1960) Heat transfer in disperse media. Nauka, Moscow

Chapter 2
Methods for the Experimental Study of the Linear Pyrolysis of Condensed Materials

Abstract Original instruments for measuring rates of linear pyrolysis (the stationary one-dimensional propagation of the reaction front in condensed compound thermal decomposition) are described. Kinetic constants of the fast high-temperature decompositions of polymers and inorganic materials (which cannot be obtained using isothermal kinetic methods) can be calculated from the experimental data provided by them. These instruments were used to study the burning mechanisms of modern solid rocket propellants over wide heating temperature ranges (up to 900 °C) and pressures (100 Pa–3 MPa). An instrument called the "chemical arc" is described for the first time. In the chemical arc, the high-temperature linear pyrolysis of solid propellant components (polymer and oxidizer) occurs in the stationary burning mode under the influence of the flat flame of interaction between the gaseous products of linear pyrolysis.

2.1 Equipment for Studying Linear Pyrolysis on a Hot Plate at Pressures P_∞=0.1 kPa–3 MPa

A schematic of the LP-1 installation used to study linear pyrolysis on a hot plate at pressures in the range 0.1 kPa–0.1 MPa is shown in Fig. 2.1.

Hot plate *1* is mounted on copper rods *2* fixed on a bulky vinyl plate. The hot plate (heater) should be made of a material (metal or alloy) with a high melting point, such as nichrome, stainless steel or niobium. In some experiments (see Sect. ??), the heater and a sample were separated by a thin layer of mica (5–10 μm) to eliminate the catalytic effect of the heater material on the kinetics of the linear pyrolysis. A Pt/Pt–Rh thermocouple (50 μm in diameter) is spot-welded to the center of the heater (just above the sample). The sample (*3*) is fixed at the end of a steel cylinder *4* which slides along a guiding Teflon rod *5*. The sample is pressed onto the heater by a load *6* which is connected to the steel cylinder (*4*) by a Capron thread that passes over a pulley fixed onto an axle. During linear pyrolysis, rotation of the pulley causes chopper *7*, mounted onto the same axle as the pulley, to rotate. A conical gear with a mirror *10* for measuring low rates of linear pyrolysis is fixed to the other end of the

Fig. 2.1 Schematic of LP-1 installation: *1*, heater; *2*, conducting rods; *3*, sample; *4*, sliding cylinder; *5*, guide rod; *6*, load; *7*, chopper; *8*, photodiode; *9*, light source; *10*, mirror

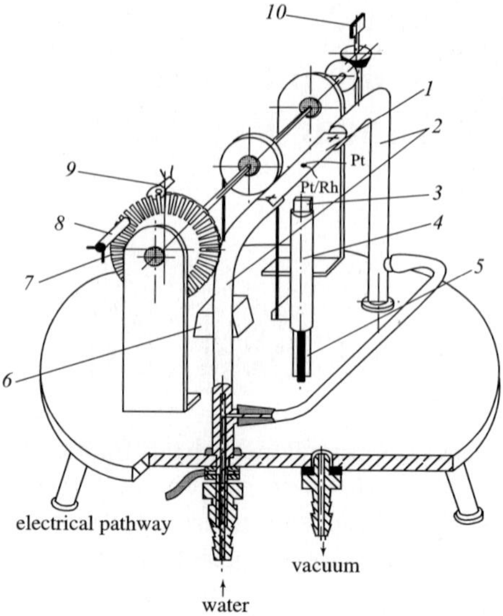

axle. As it rotates together with the axle, chopper *7* modulates a light beam incident on photodiode *8*, which is connected to a recorder.

In installations utilized in earlier studies [1, 2, 3, 4, 5], linear potentiometers were used as recorders. However the potentiometer sensors were easily damaged because they reacted with the gaseous decomposition products. Under these conditions, the accurate recording of slight displacements is impossible. The application of the installation described above allows linear pyrolysis data to be recorded with the required accuracy and reliability for systems that are interesting from various points of view and which are characterized by a very wide range of decomposition rates ($U \approx 10^{-3} - 10^{-19}\,\mathrm{m\,s^{-1}}$).

When in operation, the installation is covered by an acrylic or thick glass dome. The inner space of the installation can be evacuated or filled with inert gas through a connecting pipe (the pressure in the installation is measured by a vacuum manometer). Since the conducting rods are heated up to quite high temperatures by passing an electric current through them (up to 10^3 A) and transferring heat from heater *1*, the hollow output ends of the rods are cooled by circulating water.

Installation LP-2, which was used to measure rates of linear pyrolysis under elevated pressure, differs from LP-1 due to the presence of a constant-pressure bomb containing the heater and the kinematic unit. A schematic of installation LP-2 is shown in Fig. 2.2. Sample *7* moves downward and is pressed onto a hot plate by ring loads *8* fixed onto cylinder *9*.

Since the temperature in the top part of the bomb rises up to 80–100°C during prolonged experiments, solid-state photodiode *1* is placed in a protective water-cooled double-walled copper container. In order to improve observations of

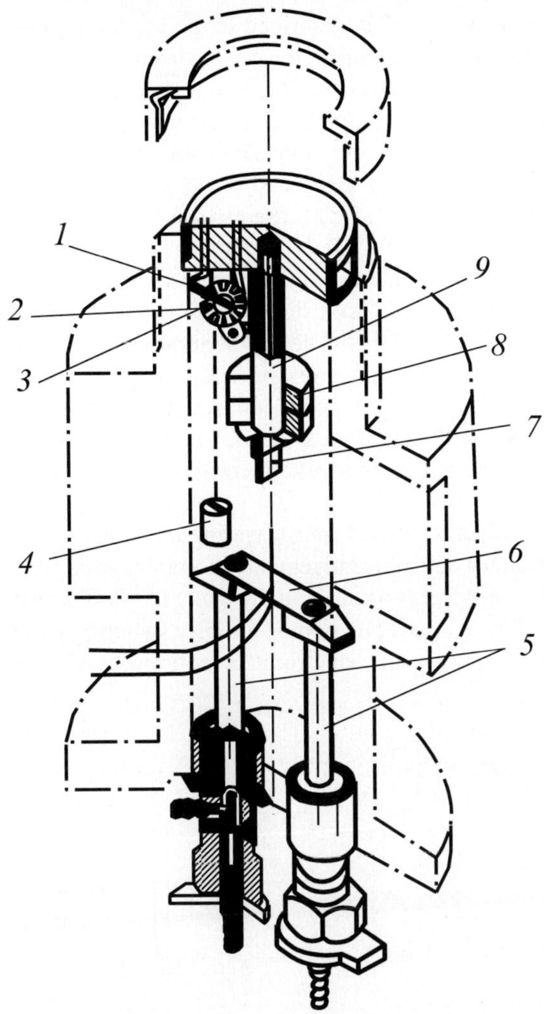

Fig. 2.2 Schematic of LP-2 installation: *1*, photodiode; *2*, chopper; *3*, light source; *4*, counterbalance; *5*, conducting rods; *6*, heater; *7*, sample; *8*, ring loads; *9*, sliding cylinder

the pyrolysis, the bomb is equipped with external lights and three windows at the level of the heater.

In both installations, the sample is separated from the heater by a pivot screen (not shown in Figs. 2.1 and 2.2) which prevents the sample from being heated before the required temperature of the heater is reached. The experiment starts at the moment at which a thin wire holding the cylinder with the sample at a distance of 25–30 mm from the heater is burnt out. The pivot screen is moved away with a spike on the cylinder as the latter approaches the heater.

In the installations for studying linear pyrolysis described in the literature [2, 3, 4, 5], the temperature was set by a voltage supplied to the heater from the outputs of a laboratory autotransformer. The main disadvantage of such a simple design is a significant decrease in the experimental data accuracy due to the instability of the heater temperature, which is caused by voltage fluctuations and convection currents in the vicinity of the heater (which are especially high at $P_\infty > 0.1\,\text{MPa}$). To avoid this problem, a point-to-point control system (Fig. 2.3) is used to stabilize the heater at its point of contact with the sample in LP-1 and LP-2.

Potentiometer PP-69, used as a setter, is connected to the thermocouple. An error signal amplified by the photogalvanic amplifier F-116 using relays R_1 (RP-4) and R_2 (MKU-48) controls the current passing through the heater by periodically engaging a ballast resistance in the input circuit of the autotransformer. The heater is supplied with power by a welding transformer T_{R2} (12 V, 10 kW) connected to the outputs of autotransformer T_{R1} (LATR-1-9A) through its primary coil. Using the described design, the temperature at the center of the heater was kept constant (±2°C) up to 850–900°C.

To conclude this section, let's consider a method for measuring the linear pyrolysis rate based on the use of noncontact optical arrangements in combination with all LP-type installations (LP-1, LP-2, LP-3, LP-4). During the study of "fast" linear pyrolysis ($U = 10^{-3}$–$10^{-5}\,\text{m}\,\text{s}^{-1}$) (Fig. 2.3, left position of tumbler switch S_1), the photodiode signal moves directly to the input of the electronic potentiometer KVT-2,

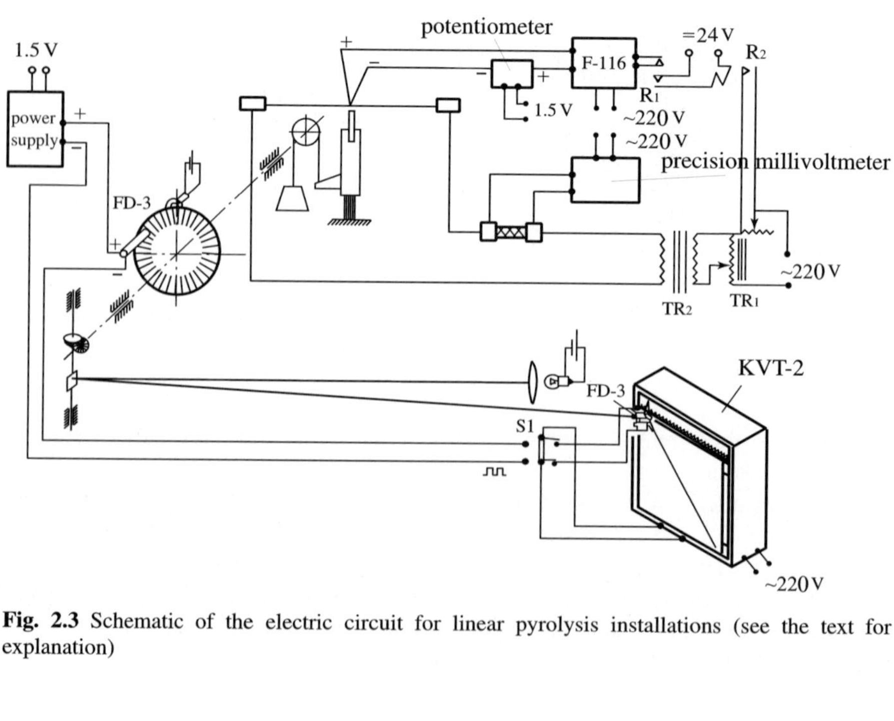

Fig. 2.3 Schematic of the electric circuit for linear pyrolysis installations (see the text for explanation)

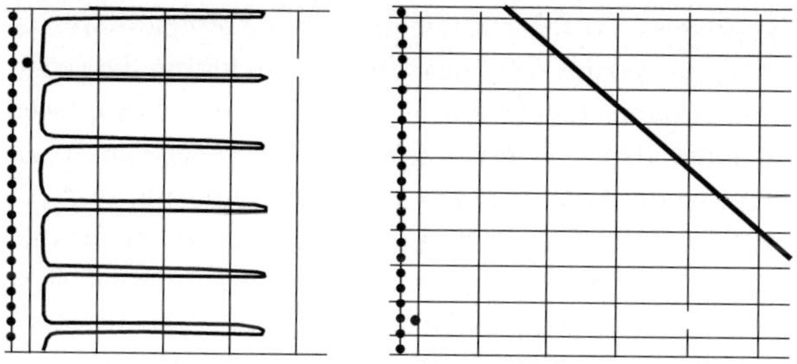

Fig. 2.4 Typical electronic recorder charts for linear pyrolysis obtained using a chopper (**a**) and a light-spot system with a photocell fixed onto the recorder slide (**b**)

which records a set of pulse signals (Fig. 2.4a). Each signal corresponds to the gap where the chopper passes between the light source and the photodiode.

The average rate of "fast" linear pyrolysis can be determined with an acceptable accuracy using a preset speed of recorder chart motion and the "chopper scale interval" (which is equal to the perimeter of the fast pulley divided by the number of chopper gaps; for LP-1, this value is 0.78–2.35 mm, and for LP-2 it is 2.15 mm). Pulse signals appearing at equal intervals indicate a steady mode of the process.

In experiments on linear pyrolysis with rates of 10^{-5}–10^{-9} m s^{-1}, choppers obviously cannot be used (the pulse signals would be separated by time intervals in the range of dozens of minutes to dozens of hours). In this case, the process parameters are recorded by a light-spot recorder. A light beam (from a standard light source) is reflected by mirror *10* (Figs. 2.1 and 2.3) to a differential photocell fixed onto an electronic potentiometer slide. The photocell consists of two photodiodes FD-3 connected in parallel. Its output signal is sent directly to the input of the recorder amplifier (in this case, the normal link between the rheochord and the recorder amplifier is disconnected).

This experimental setup is characterized by very high sensitivity and stability when recording the motion of the reflected light spot along a standard recorder chart (Fig. 2.4b). At a mirror–recorder distance of 1.2 m (LP-1, LP-3, LP-4), 1 mm of travel by the recorder pen corresponds to the decomposition of a layer of the compound 6.0 μm thick.

2.2 Equipment for Studying Linear Pyrolysis on a Bulky Heat-Conducting Block

In experiments on linear pyrolysis, the use of a thin metal plate as a heater is convenient in the case of fast decomposition. Studies of linear pyrolysis characterized by rates in the range of 10^{-6}–10^{-9} m s^{-1} are of significant interest for determining a correlation between the constants of decomposition in the volume and the macrokinetic parameters of the process. At these rates, the surface temperatures of decomposing compounds (such as composite solid propellants) are within the range near to the upper limit of the temperature region that is characteristic of processes studied by classical isothermal methods of chemical kinetics. However, due to an extremely strong (Arrhenius) temperature dependence of the decomposition rate in this region, the temperature of the heater should be kept strictly constant. In this case, a thin metal plate cannot be used as a heater because of two reasons: it is difficult to keep its temperature constant (this problem can be solved by using more sophisticated thermostating systems; for example, proportional-plus-floating control systems), and the spatially nonisothermal temperature distribution along the heating plate surface. The latter is the more serious problem, and is caused by significantly different heat exchange conditions for the parts that are in contact with the sample compared to the parts that are not. To keep the temperature constant at all points on the heater surface, one can use a bulky heat-conducting metal cylinder and press the sample onto one of its ends. The cylinder is thermostated by a nichrome heating coil wound around its lateral surface. The application of metal bodies with high heat capacity and good heat conduction properties in contact with the sample was found to be quite efficient for studying the macrokinetics of fast reactions in condensed systems (Chaps. 4–8). A photo and a schematic of the LP-3 and LP-4 installations with bulky heat-conducting heaters are shown in Figs. 2.5 and 2.6. When performing prolonged experiments on "slow" linear pyrolysis, four bulky aluminum heaters are fixed onto a single base plate for simultaneous experiments with four samples in installation LP-3. The heaters are equipped with autonomous systems of point-to-point control (Fig. 2.6) with galvanometric null indicators (bench galvanometers M-195 with built-in differential photocells).

Experimental data on the linear pyrolysis of the four samples are simultaneously recorded by an EPP-09 eight-channel recorder with a set of four vertically mounted photocells on its slide (Sect. 2.1). Preliminary alignment of the mirrors that are kinematically connected with the sample holders ensures that each light beam travels in the plane of the corresponding photocell. The cells are connected to input terminals *1, 3, 5,* and *7* via an internal switch in the recorder, while input terminals *2, 4, 6,* and *8* are shorted. Thus the slide moves along the scale to a certain spot of reflected light, where the digitized experimental point is plotted. Then the slide travels back to the beginning of the scale to move again to the next spot of reflected light, etc.

Experiments on the linear pyrolysis of doubly thermostated samples (see "Linear pyrolysis of thin samples," Sect. 1.5) were carried out using installation LP-4 (Fig. 2.6). The temperatures on the upper (T_S) and lower (T_∞) ends of a disk sample (10 mm in diameter, $\approx$ 5 mm thick) were kept strictly constant due to their contacts with a bulky heater (30 mm in diameter, 100 mm thick) and an aluminum container (*8*) filled with a heat carrier (water or glycerin pumped through by the pumping unit of a U-10 ultrathermostat), respectively. The container connected to the ultrathermostat with flexible tubing was fixed onto the upper part of the sliding cylinder, which moved on a fluoroplastic guide in the same way as in installation LP-1 (Fig. 2.1).

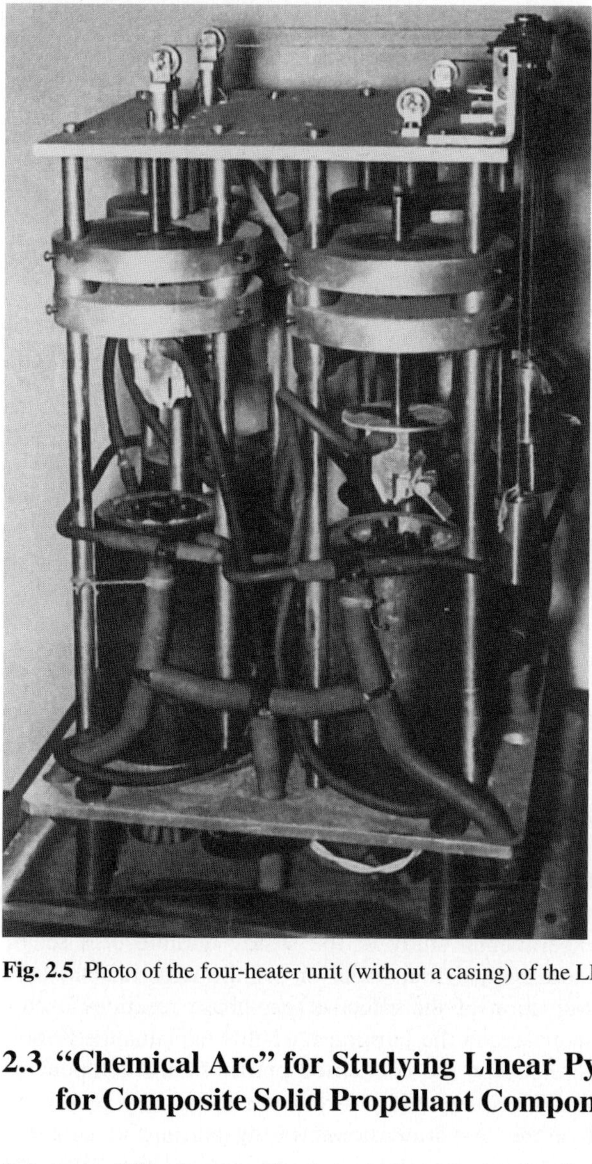

Fig. 2.5 Photo of the four-heater unit (without a casing) of the LP-3 installation

2.3 "Chemical Arc" for Studying Linear Pyrolysis Rates for Composite Solid Propellant Components in Burning Mode

It is very difficult to study the combustion mechanism of a composite solid propellant. Gasification of the condensed phase of the burning material is a complex multistage process due to the heterogeneity of the system, the different decomposition kinetics of the combustible (fuel) and the oxidizing components, as well as the influence of reactions in the condensed and gas phases on the burning rate [6, 7, 8].

To obtain more information and make conclusions about the combustion mechanisms of composite solid propellants, one needs to use special methods that allow individual stages of the process to be studied while the effects of other stages can

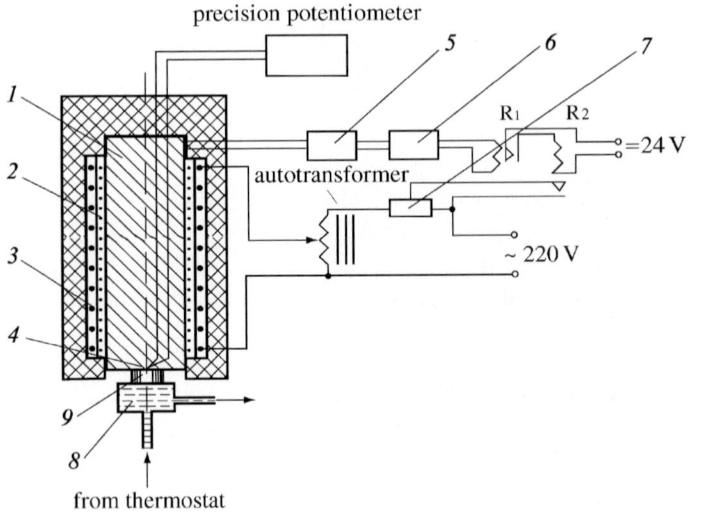

Fig. 2.6 Schematic of LP-4 installation used for pyrolysis of thin samples with double thermostating: *1*, heater unit; *2*, resistance thermometer; *3*, heating coil; *4*, thermojunction; *5*, conducting bridge; *6*, photovoltaic detector; *7*, rheostat; *8*, aluminum container; *9*, sample; R_1, R_2, relays

be eliminated, neglected or quantitatively evaluated. Methods of linear pyrolysis of samples on hot surfaces have mainly been used to study kinetics. However, they can also provide information on the macrokinetics of the individual stages of decomposition of composite solid propellant components under the conditions of purely thermal effects upon their condensed phases; in other words, without any influences from reactions between the components in the preheat zone and between products of the oxidizer and fuel gasification in the flame.

The special methods can be exemplified by the so-called sandwich technique [9, 10, 11], based on the experimental study of the steady burning of a set of alternating layers of solid oxidizer and fuel. Analysis of the experimental data results in the quantitative evaluation of the effect of gas-phase reactions occurring near the oxidizer–fuel interface on the burning rate (this explains the principal importance of the contact between the condensed phases of the components in this technique). The sandwich technique [10, 11] seems to be very promising since it represents a model of the two-dimensional steady burning of composite solid propellants. In contrast with the sandwich technique and the "chemical arc" method described below, an "end-to-end" method based on unsteady burning [2, 11] (which was confused with the chemical arc method by Strunin and coworkers [12, 13, 14, 15]) has a number of serious drawbacks. In the end-to-end experiment, burning in the contact between the ends of the pressed-together oxidizer and polymer samples is observed. In this case, the burning propagates periodically in the radial direction parallel to the contact plane, and develops in a way that is identical to that of a two-layer sandwich. At the microlevel, the burning is completely unstable, since the direction of burning front propagation changes stochastically, and various zones of the contact are always burned out. Due to the periodic nature of the contact between the condensed phases, one cannot use Cantrell's expressions (1.56) to calculate the geometry of the unsteady gas gap in the "end-to-end" technique.

In order to study some important features of the burning mechanisms of model and real materials, one should use an experimental method that applies in the absence of contact (and interaction) between the condensed phases, where linear pyrolysis occurs due to the gas flame resulting from the reaction between the products of linear pyrolysis. This is especially important when identifying the effective zone for catalysts of composite solid propellant burning. The simplest method that meets all necessary requirements is the chemical arc technique proposed by the author [16].

The chemical arc technique is quite efficient for measuring and comparing local linear pyrolysis rates under conditions of parallel burning of composite solid propellant components in the absence of contact between their condensed phases. These data can be used to determine general patterns for linear pyrolysis of composite solid propellants in the burning mode and to analyze some specific aspects of the burning mechanism (in particular, the role and "sites of action" of basic

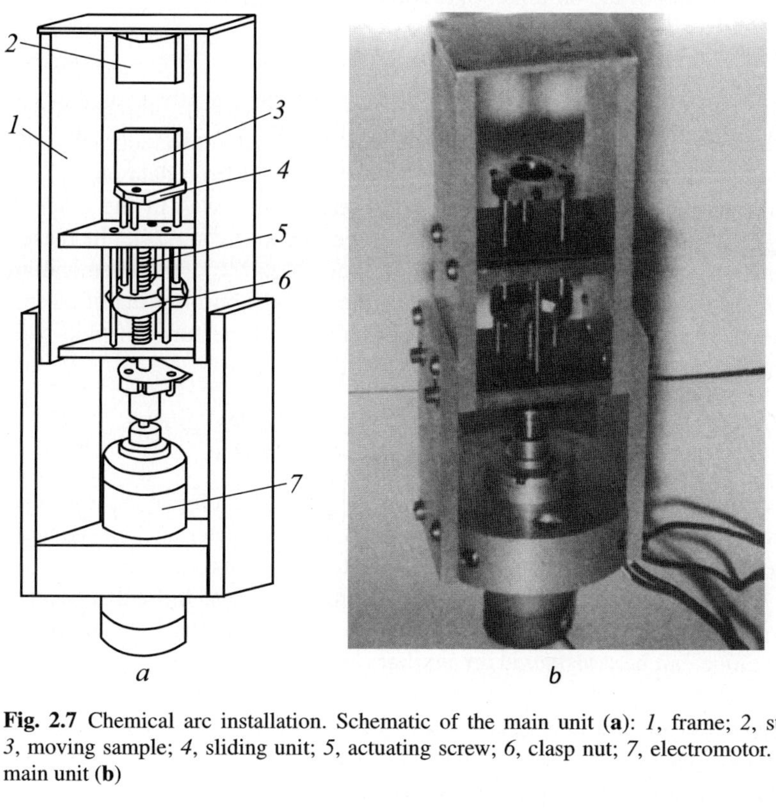

a *b*

Fig. 2.7 Chemical arc installation. Schematic of the main unit (**a**): *1*, frame; *2*, static sample; *3*, moving sample; *4*, sliding unit; *5*, actuating screw; *6*, clasp nut; *7*, electromotor. Photo of the main unit (**b**)

catalysts in composite solid propellant burning). A basic schematic diagram and a photo of the main unit of the chemical arc system are shown in Fig. 2.7. In the chemical arc, the burning of fuel and oxidizer, which play the role of the "electrodes" (by analogy to the steady burning in the high-temperature electrical plasma discharge between the electrodes in an electric arc), is maintained due to the flame produced during the interaction between the pyrolysis products of these components.

In chemical arc experiments, a fuel sample *2* (polymethyl methacrylate, PMMA, cast or pressed from a standard powder, 6 × 18 × 18 mm) was fixed in a static holder while an oxidizer sample *3* (ammonium perchlorate, AP, pressed from powders of two size fractions: 315–160 μm and 90 μm in a 1:1 ratio) of a similar size was uniformly moved towards sample *2* by an electrical drive unit.

Burning was initiated by means of a thin ballistite powder plate placed between the samples. The lateral surfaces of the samples were covered with a thin layer of nonflammable material beforehand. The velocity of the movement of the oxidizer sample, U_{mov}, was set by the pitch and rotation frequency of the actuating screw *5* (recorded by a special marker on the chart of the electronic recorder).

For steady burning of the components in the chemical arc, the width of the gap between the burning surfaces, h, is constant, and U_{mov} is related to the linear rates of burning (pyrolysis) of the oxidizer, U_{ox}, and the fuel, U_{fuel}, as

$$U_{\mathrm{mov}}(h) = U_{\mathrm{ox}}(h) + U_{\mathrm{fuel}}(h). \qquad (2.1)$$

The values of h and $U_{\mathrm{fuel}}(h)$ were recorded using a simple optical setup based on a Jupiter-3 lens system and a powerful lamp that provided a distinct shadograph of the gap on a graduated screen. Values of $U_{\mathrm{ox}}(h)$ were calculated from data on $U_{\mathrm{mov}}(h)$ and $U_{\mathrm{fuel}}(h)$ using Eq. (2.1). The steady mode setting time was in the range of 15–25 s. As U_{mov} increased, the gap width decreased, while U_{ox} and U_{fuel} increased. If the sample surfaces came into contact due to very high U_{mov}, the chemical arc phenomenon transformed into a mode involving the periodic burning-out of an end-to-end system. The experimental results were analyzed in the form of plots with coordinates:

$$U_{\mathrm{fuel}} = U_{\mathrm{fuel}}(h), \quad U_{\mathrm{ox}} = U_{\mathrm{ox}}(h), \quad \frac{\dot{m}_{\mathrm{ox}}}{\dot{m}_{\mathrm{fuel}}} = \frac{U_{\mathrm{ox}}\gamma_{\mathrm{ox}}}{U_{\mathrm{fuel}}\gamma_{\mathrm{ox}}} = f(h).$$

In contrast to methods based on regular linear pyrolysis, the chemical arc technique does not provide information on the quantitative kinetic characteristics of the high-temperature decompositions of the components. However, this method does allow one to study linear pyrolysis under conditions maximally close to those that occur in the vicinity of the burning surfaces of composite solid propellants. Thus the chemical arc technique can be considered an auxiliary method for studying linear pyrolysis in the burning mode. The application of information obtained by the chemical arc technique is considered later (Sect. 3.3), where the study of the effect of using two combustion catalysts in a composite solid propellant on the burning rates of the individual fuel components is used as an example.

References

1. Committee on Advanced Energetic Materials and Manufacturing Technologies (2004) Advanced energetic materials. Committee on Advanced Energetic Materials and Manufacturing Technologies, National Research Council of the National Academies, Washington, DC
2. Chaiken RF (1962) J Chem Phys 37:2311
3. Barsh MK, Andersen WH, Bills KW (1958) Rev Sci Instrum 29:392
4. Bills KW, Mishuck E, Moe G, Schultz RD (1959) Combust Flame 3:301
5. Chaiken RF, Andersen WH, Barsh MK (1960) J Chem Phys 32:141
6. Bakhman NN, Belyaev AF (1967) Combustion of heterogeneous condensed systems. Nauka, Moscow
7. Manelis GB, Strunin VA (1971) Combust Flame 17:69
8. Puchkov VM (1978) PhD thesis. Institute of Chemical Physics of URRS Academy of Sciences, Chernogolovka, Russia
9. Bachman NN, Librovich VB (1970) Combust Flame 15:143
10. Price EW (1993) In: 30th JANNAF Combustion Subcommittee Meeting, Monterey, CA, USA, 15–19 Nov 1993
11. Chorpening BT, Knott GM, Brewster MQ (2000) Proc Combust Symp 28:847
12. Rybanin SS, Strunin VA (1976) In: Martynenko ON (ed) Heat and mass transfer. Nauka i Tekhnika, Minsk, p 30
13. Strunin VA, Firsov AN (1977) Combust Explos Shock Waves 13:1
14. Korobeynichev OP (1971) (ed) Mechanism, kinetics and catalysis of thermal decomposition and combustion of ammonium perchlorate. Nauka, Novosibirsk
15. Manelis GB, Nazin GM, Rubtsov YI, Strunin VA (2003) Thermal decomposition and combustion of explosives and propellants. Taylor & Francis, New York
16. Shteinberg AS (1964) USSR Patent 30415

Chapter 3
Study of the High-Temperature Decomposition of AP and PMMA by the Linear Pyrolysis Method

Abstract Experimental results from studying the kinetics of the high-temperature decomposition of ammonium perchlorate (AP) and polymethylmethacrylate (PMMA) using a radically improved method of linear pyrolysis are presented. The method used allowed kinetic data corresponding to high temperatures and reaction rates to be obtained (which cannot be done using classical isothermal methods). Patterns of the high-temperature decomposition of AP and PMMA that do not manifest themselves at low temperatures are characterized. The macrokinetics of the high-temperature decomposition of AP are shown to be strongly affected by dissociative sublimation at thermodynamic equilibrium, which yields ammonia and perchloric acid vapor. The activation energies of the high-temperature decomposition of PMMA and some other polymers were found to significantly exceed those of the limiting stages of their low-temperature decomposition. As the temperature grows, the kinetic mechanism of polymer decomposition changes. Because of this, the burning of polymers occurs in the "third mode," characterized by an almost constant pyrolysis surface temperature over a wide range of polymer burning rates. Original data on the pyrolysis of solid propellant components obtained using the chemical arc (in which the stationary high-temperature decomposition of polymer and oxidizer occurs under the influence of the flat flame of interaction between the gaseous products of their linear pyrolysis) are presented. The effects of Fe_2O_3 and CuO (solid propellant combustion catalysts) on the rates of AP and PMMA linear pyrolysis are considered.

3.1 Linear Pyrolysis of AP

Despite numerous works on the combustion of composite solid propellants (for example [1, 2] and reviews collected in [3]), the ignition and burning mechanisms for these systems have been largely unclear until recently. This is explained by the lack of reliable data on the kinetics of reactions occurring in the fuel condensed phase at high temperatures corresponding to conditions in the thin reaction zone adjacent to the surface. This situation has resulted in the development of a number of speculative approaches for creating qualitative models of composite solid propellant burning. They can be exemplified by the so-called "two-temperature postulate" formulated by Schultz and Dekker [4], a theory of the thermal layer developed by Chaiken [5], and others. Well-founded ideas about the burning mechanism of composite solid propellants were formulated based on new results (including ones obtained by the author of this monograph) in recently published papers (for example [6, 7]).

Due to reasons given in the Introduction, neither data on the kinetic constants of the high-temperature decomposition of ammonium perchlorate (AP, the main basic oxidizer) and typical binders nor even qualitative data on the macrokinetics of the process have been available. The study of the burning mechanism of a composite solid propellant is an extremely challenging task, since in real systems the decomposition of the individual components occur in parallel with interaction between them (which are complicated due to the heterogeneity of the condensed phase), dispersion and heat losses from the gas phase.

Experiments described in the literature were carried out using mixtures containing model binders and AP. Most of the data available corresponds to the burning of AP–polymethyl methacrylate (PMMA) mixtures (for example [2]). For this reason, we studied the kinetics of the high-temperature decomposition of model composite solid propellants using these two compounds.

Experiments on the linear pyrolysis of AP were performed using installations LP-1, LP-2 and LP-3 and samples in the form of a cylinder (10 mm in diameter) or a plate ($18 \times 16 \times 3$ mm) pressed from a standard commercial crystal powder at heater temperatures in the range of 280–1050°C (in some experiments, samples were made from specially prepared doubly recrystallized AP. However, since at $T_0 = const$ the scattering of the linear pyrolysis rate data for standard AP samples and specially prepared ones was less than 15% given the experimental accuracy, the main sets of experiments were carried out using standard commercial AP.

Depending on T_0 (and, to a lesser degree, on the sample thickness), linear pyrolysis occurs in two qualitatively different modes: with burning at $T_0 \geq 520$–560°C, and without burning at lower T_0. At high rates of linear pyrolysis ($U \geq 10^{-6}\,\mathrm{m\,s^{-1}}$) without burning, the evolution of gaseous products (brown nitrogen oxides) was observed, while a thin layer of white AP deposit formed on cold parts of the installation (the latter formed due to the recombination of the products of equilibrium dissociative sublimation). In the experiments on very "slow" linear pyrolysis ($U < 10^{-6}\,\mathrm{m\,s^{-1}}$), the formation of the AP deposit did not take place.

First, let us consider the qualitative characteristics of AP linear pyrolysis without burning. In this case, the shape of the pyrolysis surface depends on the heater temperature, the linear pyrolysis rate and the ambient pressure. At $T_0 \leq 450$°C (heater was Al/mica, see below) and $U < 10^{-6}\,\mathrm{m\,s^{-1}}$, the pyrolysis surface is slightly concave at atmospheric pressure and at $P_\infty = 4$ kPa. At $T_0 > 480$°C and $U > 10^{-5}\,\mathrm{m\,s^{-1}}$ the pyrolysis surface becomes flat at atmospheric pressure and slightly convex at $P_\infty \ll 0.1$ MPa.

We were the first to study the catalytic effects of almost all previously used heater materials (more precisely, the oxides formed on the heater surface) on the rate of AP linear pyrolysis [8, 9]. The temperature dependencies of the rate of AP

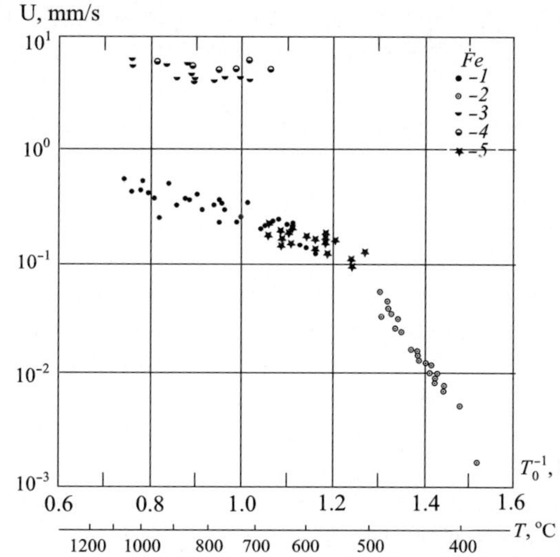

Fig. 3.1 AP linear pyrolysis on a stainless steel plate at $W/S = 0.1$ MPa and $P_\infty = 10^{-1}$ MPa with burning (*1*), without burning (*2*), for unsteady linear pyrolysis (*5*); at $P_\infty = 3$ MPa with burning (*3, 4*)

linear pyrolysis on stainless steel and platinum heaters as well as on a bulky aluminum block (LP-4) and on a thin copper or mica interlayer ($\sim 10\,\mu$m) are shown in Figs. 3.1–3.4. For samples brought into direct contact with copper, stainless steel or platinum heaters a significant increase in the decomposition rate is observed. On aluminum or mica surfaces the process occurs with much lower (commensurable) rates at the same T_0. Taking into account the lack of a noticeable influence of aluminum oxide and silicon oxide on the low-temperature decomposition of AP [3], one can conclude that the effect of aluminum and mica hot surfaces on the linear pyrolysis rate is negligibly small. In other words, the macrokinetics of pyrolysis under these conditions is determined by the kinetics of the uncatalyzed high-temperature decomposition of AP. In addition, the shape of the AP pyrolysis surface indicates that volumetric decomposition dominates, at least at $P_\infty = 0.1$ MPa and $U \leq 10^{-5}\,\mathrm{m\,s^{-1}}$.

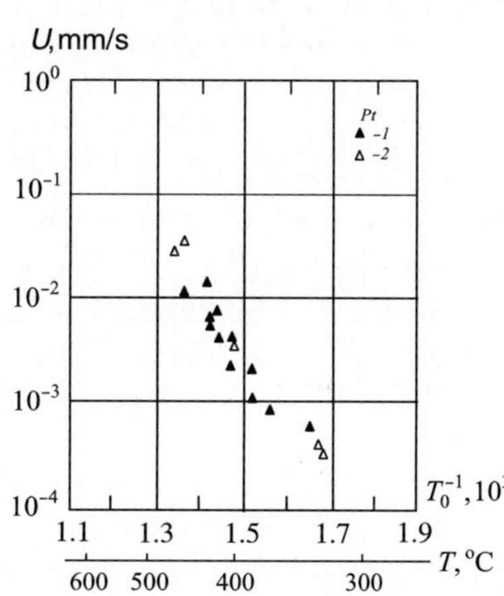

Fig. 3.2 AP linear pyrolysis on a platinum plate at $P_\infty = 0.1$ MPa and $W/S = 0.1$ MPa (*1*), 0.4 MPa (2)

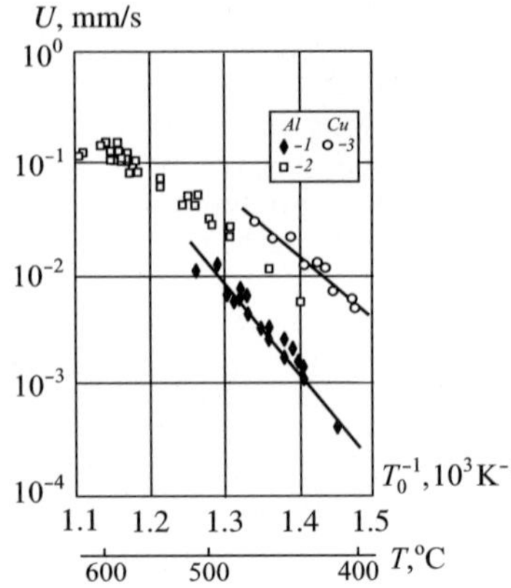

Fig. 3.3 AP linear pyrolysis on an aluminum plate at $P_\infty = 0.1\,\text{MPa}$ and $W/S = 0.1\,\text{MPa}$ (*1*), $P_\infty = 100\,\text{Pa}$ and $W/S = 0.1\,\text{MPa}$ (*2*), and on a copper plate at $P_\infty = 0.1\,\text{MPa}$ and $W/S = 0.1\,\text{MPa}$ (*3*)

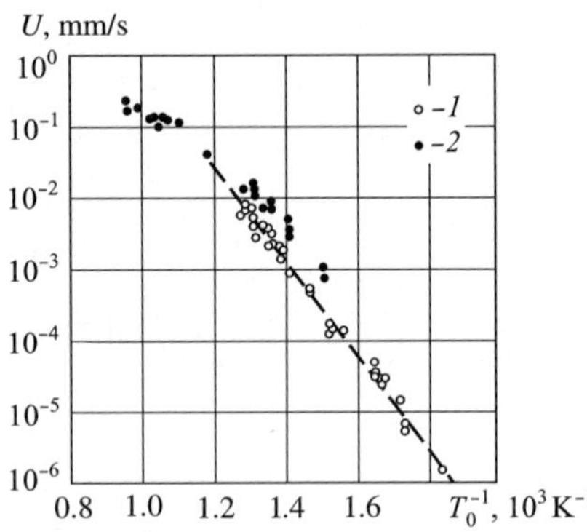

Fig. 3.4 AP linear pyrolysis on a mica interlayer at $P_\infty = 0.1\,\text{MPa}$ (*1*), 100 Pa (*2*)

Thus, the kinetic constants for the high-temperature decomposition of AP were determined from data obtained when using the mica interlayer (Fig. 3.4, curve 1). In this case, the temperature dependence of the linear pyrolysis rate can be described by the Arrhenius expression

$$U = 2.5 \times 10^6 \exp\left(-\frac{30200}{RT_S}\right) \left[\text{mm} \cdot \text{s}^{-1}\right]. \tag{3.1}$$

In the temperature range corresponding to $U = 10^{-9}$–$10^{-5}\,\text{ms}^{-1}$, the Bio number was found to be almost constant ($\text{Bi} = 0.6 \pm 0.15$ and $\dfrac{\text{M}^2}{4\text{Bi}} = 10^{-5} - 10^{-1} \ll 1$) [9]. In other words, volumetric decomposition occurs in the inner kinetic region (see Sect. 1.5), and the macrokinetics of linear pyrolysis is identical to the actual decomposition kinetics. The kinetic constants of the high-temperature decomposition of AP can be estimated using Eq. (1.37) with an accuracy that is acceptable for approximate evaluation (as shown in Sect. 8.3, the high-temperature decomposition of AP is a two-stage process; Eq. (1.37) can only be used for approximate evaluation, since the activation energies of the two stages are equal). Thus, for known values of the temperature diffusivity, thermal capacity and decomposition thermal effect

($a = 5 \times 10^{-7}\,\mathrm{m^2\,s^{-1}}$, $c = 1.46\,\mathrm{J\,g^{-1}K^{-1}}$, $Q = 1.17\,\mathrm{kJ\,g^{-1}}$), the expression for the rate constant of the high-temperature decomposition of AP in the mode of linear pyrolysis is

$$k = 5.9 \times 10^6 \exp\left(-\frac{30200}{RT}\right)\left[\mathrm{s^{-1}}\right]. \tag{3.2}$$

Let us now consider the high-temperature linear pyrolysis of AP accompanied by burning. In the case of the linear pyrolysis of a sample on aluminum and mica hot surfaces, burning was observed at $T_0 = 530$–$550°\mathrm{C}$ and $U \approx 10^{-4}\,\mathrm{m\,s^{-1}}$. For a sample on a stainless steel heater, ignition took place at approximately the same decomposition rate but at a lower heater temperature ($T_0 \approx 500$–$510°\mathrm{C}$).

If after the inflammation the sample was left motionless or was retracted from the heater, a pale yellow luminous deflagration flame was observed on the sample surface for a few seconds. The burning died out after a 1–2 mm thick layer was burnt out. After inflammation, the deflagration flame overheated the peripheral parts of the plate, which resulted in a sharp increase in the linear pyrolysis rate. However, in this mode, the temperature (T_0) dependence of the linear pyrolysis rate is much weaker than in the case of $T_0 < 500°\mathrm{C}$ (in the mode of linear pyrolysis without burning).

At the transition from T_0 to T_S, the thickness of the gas film, Z, along with the corresponding temperature drop at $U \geq 10^{-4}\,\mathrm{m\,s^{-1}}$ should be taken into consideration. However, these characteristics cannot be estimated due to the lack of reliable data on the gas film composition. The upper limiting value of T_S may be determined from experiments with a sample strongly pressed onto a heater surface (the pressure $W/S \leq 1\,\mathrm{MPa}$), which minimizes the gas film thickness. In this case, the experimental points in the plot of U against T_0^{-1} are quite scattered (a region including 36 points is shown in Fig. 3.5). However the range of T_0 (540–580°C) corresponds to the pressure of the gaseous products of the equilibrium dissociative sublima-

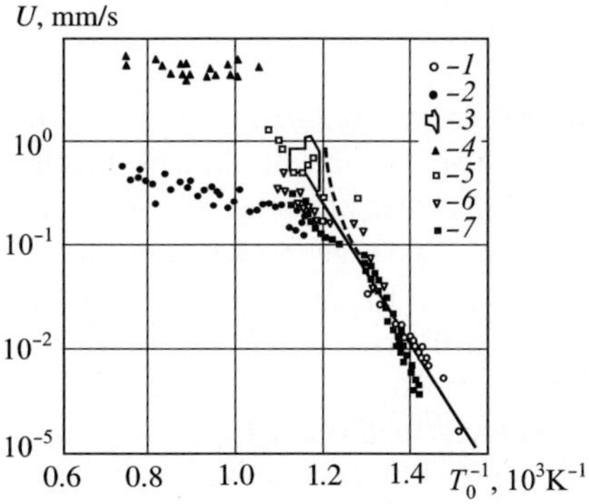

Fig. 3.5 AP linear pyrolysis on a stainless steel plate with high W/S (for the determination of T_S for AP linear pyrolysis in the burning mode) without burning at $W/S = P_\infty = 0.1\,\mathrm{MPa}$ (*1*), with burning at $W/S = P_\infty = 0.1\,\mathrm{MPa}$ (*2*), with burning at $W/S = 1.0\,\mathrm{MPa}$, $P_\infty = 0.1\,\mathrm{MPa}$ (the *outlined region* contains 36 points) (*3*), with burning at $W/S = 0.1\,\mathrm{MPa}$, $P_\infty = 3.0\,\mathrm{MPa}$ (*4*); data from [10] (*5*), [11] (*6*), [12] (*7*)

tion of AP, P_{sat} [13]. The latter is close to the average pressure in the gas film, $\overline{P} \approx P_\infty + \frac{1}{2}\frac{W}{S}$.

At high W/S, a dark (cool) spot surrounded by a luminous halo (due to the flame generated by the interaction between the gaseous pyrolysis products) indicated that endothermic sublimation plays a significant role on the sample surface. In the experiments on AP linear pyrolysis at P_∞ exceeding the atmospheric pressure, the samples inflamed and burned very rapidly. Typical experimental data obtained at $P_\infty = 3\,\mathrm{MPa}$ are presented in Fig. 3.1. In the temperature range $T_0 = 700$–$1100°\mathrm{C}$, the linear pyrolysis rate is constant [$U = (8 \pm 1) \times 10^{-3}\,\mathrm{m\,s^{-1}}$] within the scatter of the experimental points.

The mechanism of AP burning under an elevated pressure in the presence of an additional heat source is not discussed here (you can find information on this topic in [3, 6, 14, 15]). However, the AP linear pyrolysis rate (which is significantly higher than that of AP burning, $U = 40 \times 10^{-4}\,\mathrm{m\,s^{-1}}$, at $P_\infty = 3\,\mathrm{MPa}$) and its temperature dependence agree with modern ideas about the important roles of sublimation and dispersion in the high-temperature decomposition of AP in the burning mode.

Based on the data presented above, one can estimate the deep decomposition constant of AP in the temperature range of $T_S = 280$–$500°\mathrm{C}$ using Eq. (3.2). At higher temperatures ($T_S \geq 550°\mathrm{C}$), the dominant role in the pyrolysis macrokinetics under the conditions of unhindered gas removal is played by dissociative sublimation (which yields NH_3 and $HClO_4$) [13]. Reactions between the products of dissociative sublimation above the AP surface lead to the appearance of a deflagration flame. This conclusion is in excellent agreement with the results from experiments on the linear pyrolysis of thin AP samples carried out using installation LP-4 [16].

Some features of the high-temperature decomposition of AP are also considered below, in Sects. 3.3 and 3.4, and in Chaps. 8 and 9.

3.2 Linear Pyrolysis of PMMA

Due to the wide application of polymeric materials in engineering, the kinetics of their thermal destruction has been studied in detail. So far, the kinetic mechanisms for low-temperature destruction in the volume that occur at relatively low rates (half-lives in the range from minutes to hours) have been characterized for polymers from various classes (see for example [17, 18, 19]). The application of physical methods (calorimetry, thermogravimetry, manometry and others) in combination with chromatography and mass-spectrometry has allowed a relationship between the destruction rates and the structures of the corresponding polymeric systems to be established. The accompanying polymer thermal destruction chain-radical reactions have been thoroughly investigated. As a result of this research, a number of polymeric materials with enhanced service characteristics (thermal stability, radiation hardness, resistance to oxidative aging, etc.) have been developed.

However, the forced high-temperature decomposition of polymeric materials has been studied to a much lesser degree. This can be explained by some methodological difficulties (see Introduction) as well as by the novelty of this research area, which is mainly connected with space engineering. In this situation, the study of gross-process macrokinetics is more important than the characterization of reaction mechanisms for complex transformations of a polymer. Since polymers are used as components of composite solid and hybrid propellants, studies of the effects of high-temperature destruction macrokinetic parameters on polymer burning patterns are of special interest.

In this case, the redundancy of data on the low-temperature kinetics of polymer decomposition can be illustrated by the following example. At $T \leq 350°C$, PMMA decomposition can be described as a first-order reaction (at $\eta > 0.2$) with activation energy $E \approx 165\,\mathrm{kJ\,mol^{-1}}$. By using corresponding expressions for the rate constant and Eq. (1.1), one can obtain the temperatures of the polymer surface, $T_S = 800$ and 900°C, for burning rates in the range of 10^{-3}–$10^{-2}\,\mathrm{m\,s^{-1}}$ (since Eq. (1.1) is valid for decomposition due to a zero-order reaction, the temperature values obtained are the lower bound ones). In both cases, the estimated temperature values significantly exceed the real ones determined from, for example, microthermocouple measurements [20]. The application of the extrapolation method in order to estimate the surface temperature using kinetic equations of thermal decomposition [1, 21, 22, 23] leads to a similar result. This is because the influence of the effect of the gas-film on the accuracy of T_S determination is neglected.

Thus, studies of the macrokinetics of the forced thermal decomposition of polymers in the linear pyrolysis mode using improved experimental and theoretical techniques were of primary importance. In the experiments, samples of linear (thermoplastic) and crosslinked (thermosetting) PMMA were investigated.

Pieces of standard Plexiglas were used as the linear PMMA samples, while crosslinked PMMA samples were specially prepared by the block polymerization of methyl methacrylate with the addition of 2 or 10% triethyleneglycoldimethacrylate (crosslinking agent). The crosslinked PMMA samples did not dissolve in acetone and dichloroethane and did not exhibit softening or melting during pyrolysis.

3.2.1 Linear Pyrolysis of Thin ("Finite Length") PMMA Samples Under Double Thermostating

As was mentioned in Sect. 1.5, the mechanism of decomposition (volumetric or on the surface) can be directly determined by linear pyrolysis experiments with "finite length" samples under double thermostating of the hot and cold surfaces. For the volumetric mechanism, the linear pyrolysis rate should decrease exponentially with time, while for the surface mechanism, $U = \mathrm{const} \neq f(t)$. For crystalline samples (for example AP or salmiac), the decomposition mechanism can be identified from the shape of the pyrolysis surface: concave in the case of the volumetric reaction of AP (at $T_S < 500°C$), or convex in the case of the dissociative sublimation of salmac at the surface.

In contrast to these inorganic salts, the shape of the high-temperature pyrolysis surface of the polymer (especially a thermoplastic one) can depend not only on the kinetic parameters of the process but also on the physicomechanical characteristics of the compound. Therefore, the macrokinetic mechanism of high-temperature polymer decomposition should be determined based on the parameters of the unsteady linear pyrolysis of "finite length" samples.

The experiments were performed using installation LP-4 and PMMA samples in the form of thin discs (diameter $d = 10^{-2}$ m, length $l_0 = (1.6–1.8) \times 10^{-3}$ m) at $W = 10.3$ N. The pyrolysis was stopped as soon as the sample length attained $l = (0.4–0.5) \times 10^{-3}$ m, since further sample thinning meant that the temperatures at its hot and cold ends could not be kept constant with the required accuracy.

Some typical plots of the sample length against time for crosslinked PMMA (10% crosslinking agent) are shown in Fig. 3.6. Since the pressing force does not influence the linear pyrolysis rate at $T_0 \leq 480°\text{C}$, the temperature drop across the gas film can be neglected ($T_0 = T_S$). The $l(t)$ dependencies (Fig. 3.6a) can be satisfactorily linearized in semi-logarithmic coordinates, which indicates the volumetric mechanism of the polymer decomposition. The kinetic constants of the overall limiting reaction can be estimated using Eqs. (1.42) and (1.43) describing the linear pyrolysis of "finite length" samples. The corresponding relationships are

$$l = l_0 \exp(-t \cdot tg\varphi), \tag{3.3}$$

$$tg\varphi = \frac{RT_S^2 k_0 \exp(-E/RT_S)}{E(T_S - T_\infty)}, \tag{3.4}$$

where φ is the slope of the line in the semi-logarithmic coordinates (Fig. 3.6b).

Analysis of the experimental data (partially shown in Fig. 3.6) in Arrhenius coordinates using Eqs. (3.3) and (3.4) is illustrated in Fig. 3.7. The resulting expression for the rate constant of crosslinked PMMA decomposition is

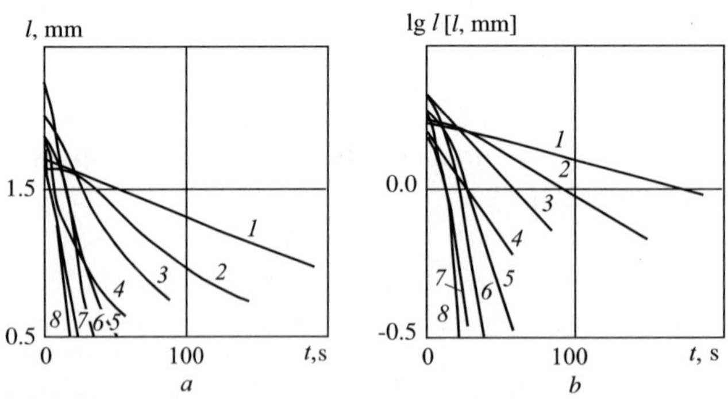

Fig. 3.6 Linear pyrolysis of crosslinked PMMA (installation LP-4, thin samples under double thermostating): (**a**) sample thickness versus time at $T_\infty = 20°\text{C}$ and $T_S = 389°\text{C}$ (*1*), 410°C (*2*), 424°C (*3*), 431°C (*4*), 444°C (*5*), 463°C (*6*), 468°C (*7*), 481°C (*8*); (**b**) the same dependencies in semilogarithmic coordinates

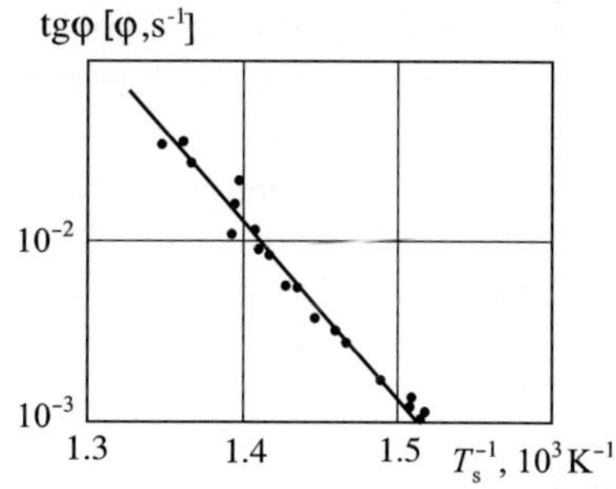

Fig. 3.7 Kinetics of the high-temperature decomposition of crosslinked PMMA [analysis of data shown in Fig. 3.6 by using expression (3.4)]

$$k = 10^{12.5\pm1.0}\exp\left(-\frac{43000\pm3000}{RT}\right)\left[s^{-1}\right]. \quad (3.5)$$

3.2.2 *Linear Pyrolysis of Cylindrical PMMA Samples in the Presence of a Gas Film*

In the experiments with samples of linear and crosslinked PMMA ($l = 50$ mm, $d = 6$–12 mm), installations LP-1, LP-2 and LP-4 were used. As well as the heater temperature, T_0, the sample diameter, b, and the pressing force, W, were varied in the following ranges: $b = (0.6$–$1.2) \times 10^{-2}$ m, $W = 5$–17 N.

The experimental data are shown in Fig. 3.8 as a plot of lg$U$ against T_0^{-1} (a coordinate system used in previous works on linear pyrolysis). In the studied range of T_0, Arrhenius linearization does not take place, and the degree of curve divergence increases with increasing T_0 and U. This agrees qualitatively with ideas regarding the gas-film effect: the temperature drop across the gas film, $\Delta T = T_0 - T_S$, and its thickness, Z, increase with increasing U and b and decreasing W.

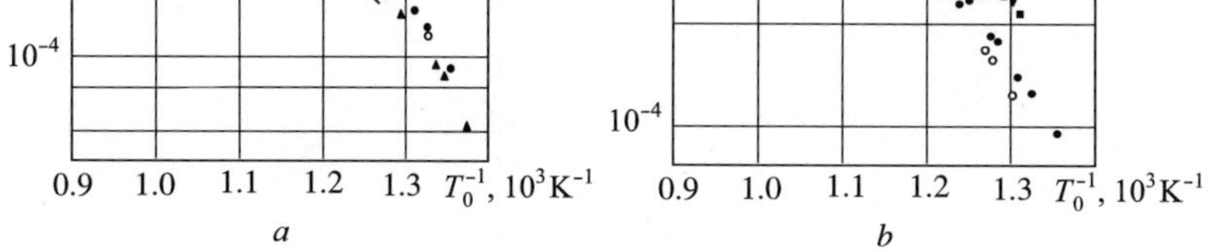

Fig. 3.8 The influence of the sample diameter, b, and the pressing force, W, on the linear pyrolysis of cylindrical samples of linear PMMA: (**a**) at $b = 1.2 \times 10^{-2}$ m and $W = 16.5$ N (*1*), 12.3 N (2) (points 2' correspond to the niobium plate), 9.9 N (*3*), 5.5 N (*4*); (**b**) at $W = 16.5$ N and $b = 1.2 \times 10^{-2}$ m (*1*), 1.0×10^{-2} m (2), 0.8×10^{-2} m (*3*), 0.6×10^{-2} m (*4*)

Let's analyze the data obtained from the viewpoint of flow hydrodynamics in the gas gap. In the theory developed by Cantrell, the influence of sliding on the pyrolysis surface on the gap parameters is assumed to be negligible. For crosslinked polymers (as well as for solid carbonic acid, AP and salmiac) this assumption is correct, while for linear PMMA special verification is required.

According to the results obtained in [24], at $T_0 = \text{const}$

$$U \sim b^{-n} \tag{3.6}$$

at $W = \text{const}$ and

$$U \sim W^{m} \tag{3.7}$$

at $b = \text{const}$, where $n = 1$ and $m = 0.25$.

Analysis of the experimental data (Fig. 3.8) using the method illustrated by Figs. 3.9 and 3.10 results in $n = 0.96$ and $m = 0.27$, which is in a very good agreement with theoretical values [25].

Due to the good agreement between the experimental results and theoretical relationships obtained upon analyzing the hydrodynamics in the gas gap, the dependence of the decomposition rate on the material surface temperature can be determined using the approximate method described in Sect. 1.7. Data on the linear pyrolysis of linear and crosslinked PMMA were plotted in $(\dot{m}b)^{4/3}W^{-1/3}$–T_0 coordinates (Fig. 3.11). The thermal conductivity of the gaseous products of PMMA decomposition was assumed to be almost independent of the temperature. Upon reaching a certain value (520°C for linear PMMA and 600°C for crosslinked PMMA), T_S does not markedly increase as the heat-flux density and the linear pyrolysis rate increase. Under these conditions, the temperature drop across the gas gap is very high.

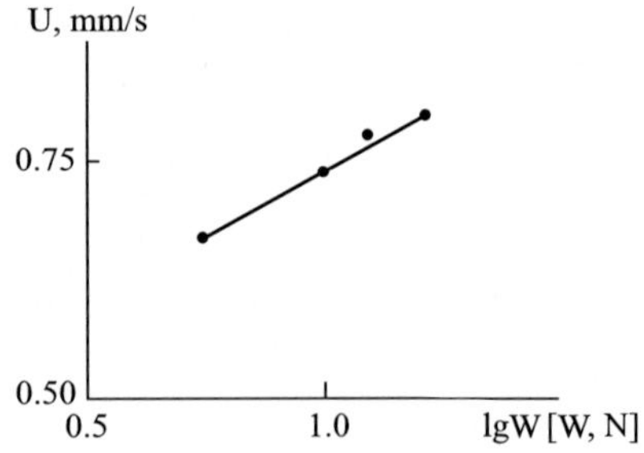

Fig. 3.9 Determination of m in expression (3.7) from data shown in Fig. 3.8a

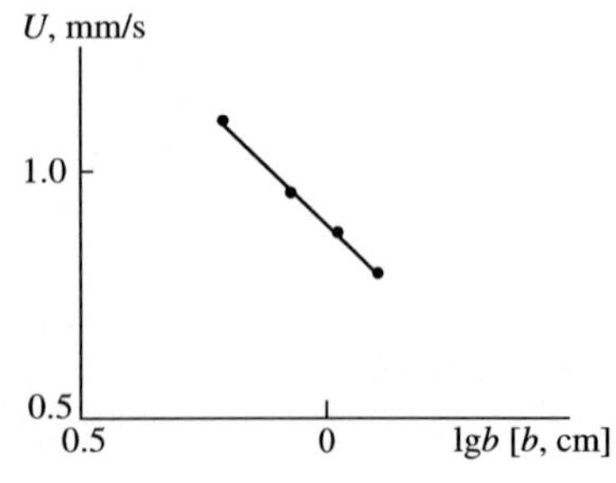

Fig. 3.10 Determination of n in expression (3.6) from data shown in Fig. 3.8b

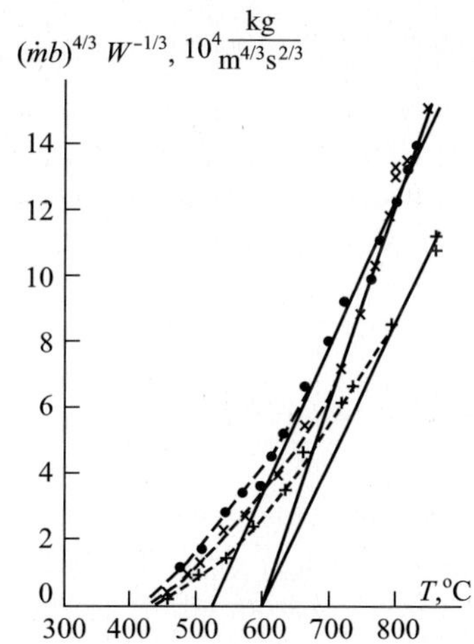

Fig. 3.11 Determination of $U\,(T_S)$ from data on $U\,(T_0)$ for linear PMMA (*circles*) and for samples of crosslinked PMMA with 2% (*multiplication symbols*) and 10% (*crosses*) crosslinking agent

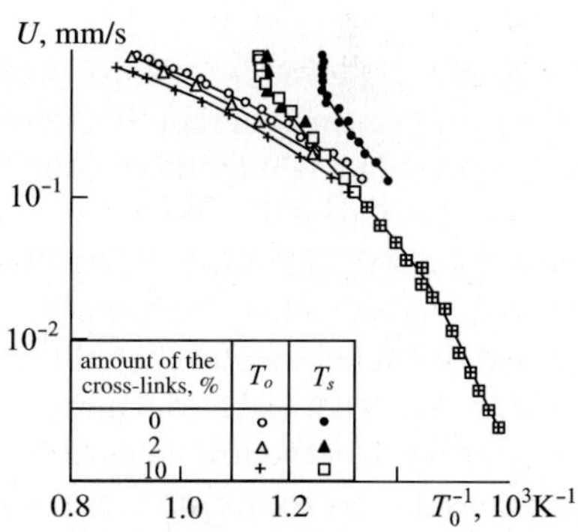

Fig. 3.12 Experimental relations for $U(T_0)$ and $U(T_S)$, calculated using the approximate method for the linear pyrolysis of linear and crosslinked PMMA

Data on the linear pyrolysis of linear and crosslinked PMMA are presented in the Arrhenius anamorphism in Fig. 3.12 (the T_S values are determined from the data presented in Figs. 3.8 and 3.11 using the approximate method described above). The $U(T_0)$ dependencies are also shown in Fig. 3.12 for comparison.

3.2.3 Linear Pyrolysis of PMMA at Elevated Pressures

In the experiments with flat samples of PMMA ($3 \times 18 \times 30$ mm) at $W = 4$ N, $P_\infty = 2$ and 5 MPa (N_2), $T_0 = 450$–700°C and $U = 0.1$–$0.6\,\mathrm{mm\,s^{-1}}$, installation LP-2 was used. In order to be able to compare the results, similar experiments were performed at a normal pressure.

At $P_\infty \leq 2$–3 MPa and $U = 0.1$–$0.6\,\mathrm{mm\,s^{-1}}$, steady linear pyrolysis was observed. The steady state of the process under the indicated conditions was also confirmed by special experiments performed at $P_\infty = 1$, 3, 4 and 6 MPa.

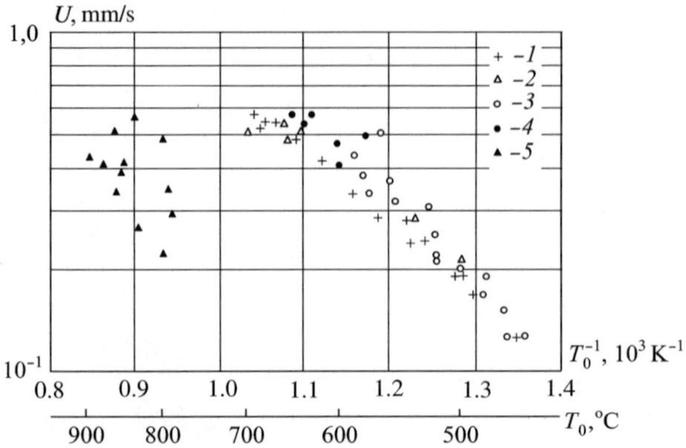

Fig. 3.13 Linear pyrolysis of PMMA at $P_\infty = 0.1$ MPa (*1*), 4.0 MPa (2), 5.0 MPa (no coke) (*3*), 5.0 MPa (coke) (*4*), 4.0 MPa (coke) (*5*)

At $P_\infty \geq 4$ MPa, $T_0 \geq 600°$C and $U \geq 0.5$ mm s^{-1}, a layer of a condensed coke-like pyrolysis product rapidly grows on the heater surface, resulting in fast overheating of the plate. Under these conditions, the decomposition rate decreases (despite the overheating of the plate) due to the increase in the thermal resistance between the heater and the sample. At $P_\infty = 4$ MPa, the coke formation starts at higher T_0 and U (640°C and 0.6 mm s^{-1}) than at $P_\infty = 5$ MPa (600°C and 0.5 mm s^{-1}, respectively). Based on these data, one could suggest that the coke formation would be observed at $P_\infty \approx 2$–3 MPa and $U \sim 1$ mm s^{-1}. However, this could not be verified experimentally because of the low capacity of the power transformer and the small volume of the constant-pressure bomb (the latter resulted in a disturbance of the steady process due to the heating of the sample with hot bomb-filling gas at high T_0).

The experimental data are shown in Fig. 3.13 in the primary coordinates $\lg U - T_0^{-1}$.

At $T_0 =$ const, the increase in the external pressure almost does not influence the linear pyrolysis rate. At $U =$ const, the temperature of the heater is lower at $P_\infty =$ 0.1 MPa (Fig. 3.13, points 1) than at $P_\infty = 4$ MPa (Fig. 3.13, points 2). However, this can be attributed to the gas-film effect, since the thickness of the gas film $Z \sim P_\infty^{-1/3}$ [23]. Thus, for PMMA, the dependence of U on T_S is hardly affected by changes in P_∞ (at least over the range studied).

3.2.4 *Analysis of the Kinetics of PMMA High-Temperature Decomposition Based on Linear Pyrolysis Data*

Before discussing the results obtained, a general remark regarding methods for studying the linear pyrolysis of polymers should be made.

A set of experimental data on the pyrolysis of crosslinked PMMA (which does not soften or melt during decomposition under experimental conditions) is the most useful information for analyzing linear pyrolysis kinetics over wide ranges of T_0 and U. The expressions of linear pyrolysis theory can be used to analyze the data covering the full temperature and the decomposition rate ranges.

For fast linear pyrolysis ($U > 0.1\,\mathrm{mm\,s^{-1}}$), the shapes of the thermoplastic polymeric samples do not change due to the small thickness of the preheat zone and the high viscosity of the melt (the characteristic time of the "pressing-out" of the softened material is much higher than its pyrolysis time).

For slow linear pyrolysis ($U < 0.1\,\mathrm{mm\,s^{-1}}$), the thickness of the preheat layer in the PMMA sample becomes comparable with its size (diameter). Under these conditions, the shapes of crosslinked PMMA samples do not change, while for linear PMMA samples, expansion of the hot end is observed when decomposition begins (the change in the sample shape remains constant during the linear pyrolysis process). This phenomenon was ignored in previous studies of linear pyrolysis.

Thus, data on the fast linear pyrolysis of linear PMMA (that is, for a constant sample shape throughout the process) should be used for quantitative analysis.

Let us consider the dependence of the crosslinked PMMA decomposition rate on the surface temperature (square points in Fig. 3.12). Three regions characterized by different dependence patterns can be distinguished.

In region I ($T_S < 700\,\mathrm{K}$, $U < 3\times10^{-2}\,\mathrm{mm\,s^{-1}}$), the relationship between the decomposition rate and the surface temperature can be described as $U \sim \left(-\frac{43000}{RT_S}\right)$. In region II ($T_S = 700$–$860\,\mathrm{K}$, $U = 0.03$–$0.7\,\mathrm{mm\,s^{-1}}$), the slope of the straight line $\lg U = \lg U_0 - A/T_S$ decreases by approximately a factor of two: $U \sim \left(-\frac{21000}{RT_S}\right)$. In region III, the increase in U caused by the growth in the heater-generated heat flux density is not accompanied by a noticeable increase in the pyrolysis surface temperature.

Since PMMA decomposition occurs in the volume (see Sect. 3.2.1), the expressions derived in Sect. 1.5 can be used to analyze the fast linear pyrolysis of semi-infinite samples. For cylindrical samples ($d = (6$–$8)\times10^{-3}\,\mathrm{m}$) at U and T_S characteristic of regions I and II, the Bio number was found from direct heat-flux density measurement data [25] to be 0.3–0.4. According to Eq. (1.32), the transfer of the linear pyrolysis macrokinetics from the inner kinetic mode to the inner diffusion mode should take place at $U \approx 4\times10^{-2}\,\mathrm{mm\,s^{-1}}$, which is in a good agreement with the boundary between regions I and II (Fig. 3.12).

Analysis of the data shown in Fig. 3.12 using Eqs. (1.1) and (1.37) resulted in the following values for the volumetric decomposition rate constant:

$$k = 3.3\times10^{12}\exp\left(-\frac{43000}{RT}\right)\left[s^{-1}\right] \text{ for crosslinked PMMA}$$

($650 < T_S < 700\,\mathrm{K}$, the inner kinetic mode; $700 < T_S < 860\,\mathrm{K}$, the inner diffusion mode) and

$k = 2.1 \times 10^{13} \exp\left(-\frac{43000}{RT}\right) \left[s^{-1}\right]$ for linear (thermoplastic) PMMA
($700 < T_S < 790\,\mathrm{K}$, the inner diffusion mode).

The rate constants obtained by the methods of linear pyrolysis of semi-infinite samples and thin films under double thermostating were found to be in a good agreement.

In conclusion, let's consider the results corresponding to region III (the so-called *third mode* of linear pyrolysis). The effective activation energy for the process (E_{eff}) in this region cannot be determined by existing experimental methods based on linear pyrolysis. However, approximate evaluation for linear and crosslinked PMMA yields $E_{eff} \geq 290$–$330\,\mathrm{kJ\,mol^{-1}}$, provided that volumetric decomposition of PMMA occurs in region III. In this case, the preexponential factors are also believed to be quite high ($k_0 \geq 10^{20}$–10^{24} s).

One can suppose that these constants may correspond to a real process occurring in the polymer and may reflect the existence of a stage of mainly polymer bond breaking (among the entire set of parallel reactions). This randomly occurring process can be characterized by $E_{eff} = 290$–$330\,\mathrm{kJ\,mol^{-1}}$, which is very close to the C–C bond energy (290–330 $\mathrm{kJ\,mol^{-1}}$) [17]. It is reasonable that in the presence of parallel reactions with lower activation energies ($E \sim 165\,\mathrm{kJ\,mol^{-1}}$), this stage only manifests itself at higher temperatures, in region III (the third mode of linear pyrolysis).

The additional reason for the appearance of the third mode is the occurrence of dispersion at high temperatures. As was mentioned above (see Chap. 1, Fig. 1.5), dispersion should lead to a dramatic acceleration of the propagation of the front for a slight increase in the surface temperature. The content of dispersed particles in the products of PMMA pyrolysis has not been determined yet, though the dispersion phenomenon accompanying PMMA linear pyrolysis and burning has been observed in experiments [26]. The weak dependence of the total heat of polymer decomposition, Q_Σ (see Chap. 4) on the linear pyrolysis rate (at least at $U \sim 1\,\mathrm{mm\,s^{-1}}$) may indicate that a change in the kinetic mechanism is an important cause of the appearance of the third mode.

Results from studying the kinetics and products of methyl methacrylate monomer decomposition [27] also suggest a change in the PMMA thermal decomposition mechanism at temperatures of over 520°C. The rates of methyl methacrylate decomposition were found to be several orders of magnitude lower than those of PMMA. Thus, the abrupt change in the composition of PMMA linear pyrolysis products can be explained by a change in the initial polymer decomposition mechanism rather than by secondary reactions of gaseous monomer decomposition.

Data on the constants for the high-temperature decomposition of PMMA, as determined by linear pyrolysis-based methods, are summarized in Table 3.1. Some literature data on kinetic studies of the low-temperature decomposition of PMMA performed by classical isothermal techniques are given in Table 3.2 for comparison.

Table 3.1 Kinetic constants of PMMA decomposition determined from linear pyrolysis data

Polymer	T, °C	E, kJ mol^{-1}	lgk_0 [k_0, s^{-1}]
PMMA	450–520 >530	180 ± 12 ≥ 290–330	13.3 ± 1.0 22–23
PMMA + 2 wt% triethyleneglycoldimethacrylate	400–590 >600	180 ± 12 ≥ 290–330	12.5 ± 1.0 21–22
PMMA + 10 wt% triethyleneglycoldimethacrylate	400–590 >600	180 ± 12 ≥ 290–330	12.5 ± 1.0 21–22

Table 3.2 Literature data on the kinetic constants of low-temperature PMMA decomposition

	Polymer	T, °C	E, kJ mol^{-1}	lgk_0 [k_0, s^{-1}]	Reference
1	Linear PMMA	220–250	130	9.5	[17]
2		226–256	125	8.6	[18]
3		290–340	177	12.5	[26]
4		380–430	180	13.0	[28]
5	PMMA + 2 wt% triethyleneglycoldimethacrylate	310–360	181	12.1	[26]
6	PMMA + 10 wt% triethyleneglycoldimethacrylate	310–360	182	12.3	[26]

3.3 Linear Pyrolysis of AP and PMMA in the Chemical Arc: Effect of Combustion Catalysts

The use of the chemical arc installation allows one to study the joint linear decomposition of fuel and oxidizer in the presence of the interaction flame. These conditions are similar to those occurring in the vicinity of the burning surfaces of composite solid propellants.

In this section, experimental data on the linear pyrolysis of AP and PMMA in the chemical arc are discussed. Samples made from pure components as well as catalyst-containing (Fe_2O_3 or CuO) ones were studied. The influence of Fe_2O_3 and CuO on AP and PMMA burning in the chemical arc attracted the author's attention for the first time in 1963 [29]. This aspect was later investigated in detail in collaboration with Benin and Rodin. The main results of those experiments are summarized in this section.

The experimental data on the rates of pure AP and PMMA burning in the chemical arc are shown in Fig. 3.14. The ratio of the weight rates of PMMA and AP burning ($\xi = 0.30 \pm 0.04$) exhibits a weak dependence on the width of the gap between the pyrolysis surfaces, h, within the range of 0.5–4 mm. The introduction of Fe_2O_3 to AP (at concentrations as high as 4 wt%) does not affect the burning rates of the components. In the chemical arc, the bright flame of interaction between the pyrolysis products was found to be much closer to the surface of the fuel than that of the oxidizer. At low velocities of the oxidizer sample moving (u_{mov}) and

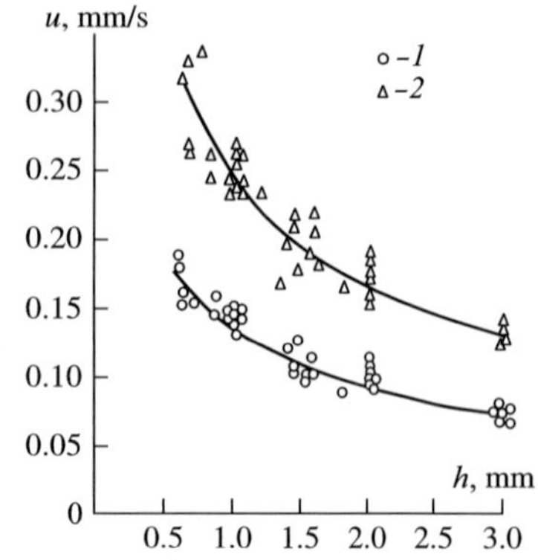

Fig. 3.14 Rates of catalyst-free PMMA (*1*) and AP (*2*) burning (linear pyrolysis) in the chemical arc versus the width of the gap between the pyrolysis surfaces

correspondingly at rather wide gaps, a pale flame of AP deflagration can be clearly seen near the AP surface (Fig. 3.15).

Introduction of 4 wt% CuO into AP results in a dramatic increase in the burning rates for both components (Fig. 3.16). At $h < 1.5$ mm, the ratio of the weight rates of burning is still constant ($\xi \approx 0.14$). As the gap width is further decreased, a relative increase in the linear pyrolysis rate for the fuel is observed (at $h \approx 0.1$ mm $\xi \approx 0.3$). At 1% CuO in AP, the curves of $U_{\rm ox}(h)$ and $U_{\rm fuel}(h)$ are cymbate, and the ratio of the weight rates remains constant and equal to 0.21 ± 0.02 (Fig. 3.17). For the (PMMA + 5% CuO)–(AP) system, burning in the chemical arc is quantitatively identical to that in the catalyst-free system. The introduction of Fe_2O_3 into PMMA leads to an increase in its burning rate in comparison with the pure PMMA (Fig. 3.18), while the rate of AP burning is lower than that of the catalyst-free system. For the (PMMA + 1% Fe_2O_3)–(AP) system, ξ is equal to 0.44 ± 0.03 in the gap width range indicated above. The increase in the Fe_2O_3 content in PMMA from 5 to 10% does not influence the efficiency of the catalyst. The data obtained for PMMA–AP system burning are summarized in Table 3.3.

Based on the data obtained, some conclusions regarding the mechanism for the catalytic burning of the model solid propellant can be made. The qualitative scheme for transformations of the components while burning in the chemical arc is as follows. The fuel (PMMA) gasification occurs via its linear decomposition due to the

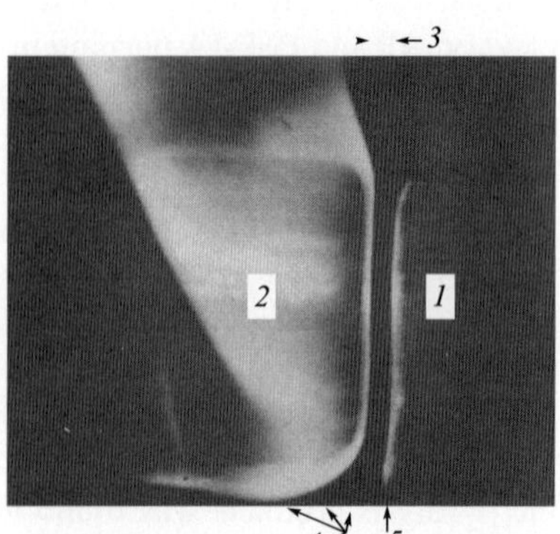

Fig. 3.15 Photo of AP (*1*) and PMMA (*2*) steady burning in the chemical arc (*3*, the gas gap; *4*, diffusion flame from interaction between the gaseous products of PMMA linear pyrolysis and the products of the AP deflagration flame, *5*)

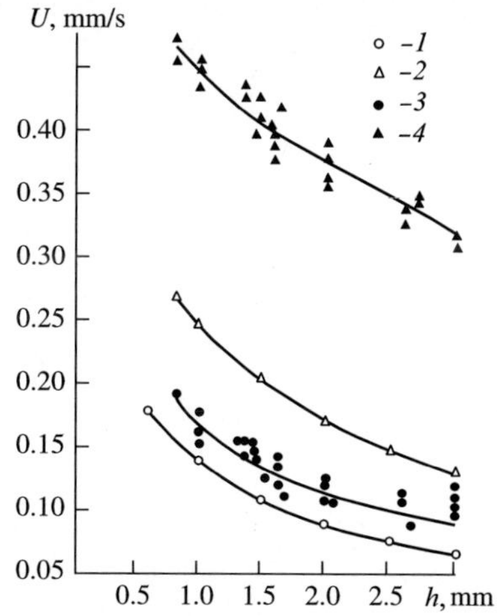

Fig. 3.16 Rate of linear pyrolysis (burning) of the components of a model solid propellant in the chemical arc versus the width of the gap between the pyrolysis surfaces for the PMMA (*1*)–AP (*2*) system and for the PMMA (*3*)– AP + 4 wt% CuO (*4*) system

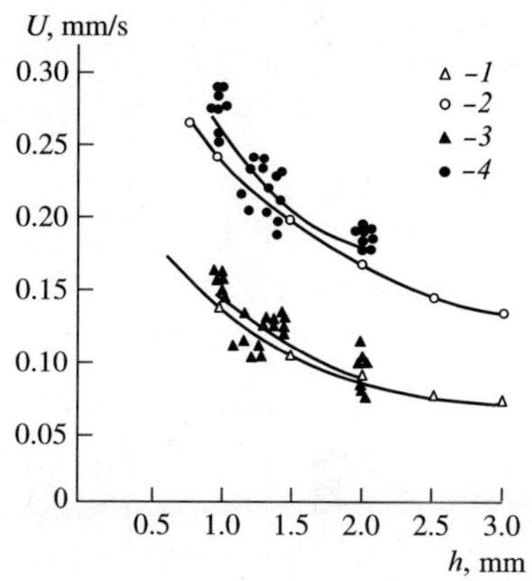

Fig. 3.17 Rate of linear pyrolysis in the chemical arc versus the width of the gap between the pyrolysis surfaces for the PMMA (*1*)–AP (*2*) system and for the PMMA (*3*)– AP + 1 wt% CuO (*4*) system

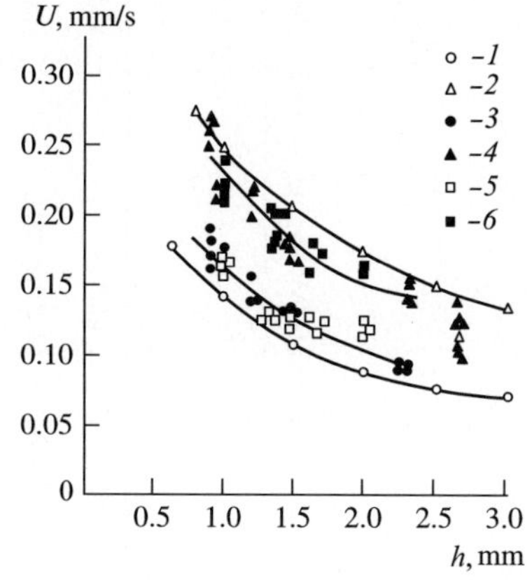

Fig. 3.18 Rate of linear pyrolysis in the chemical arc versus the width of the gap between the pyrolysis surfaces for the PMMA (*1*)–AP (*2*) system, for the PMMA + 5 wt% Fe_2O_3 (*3*)– AP (*4*) system, and for the PMMA + 10 wt% Fe_2O_3 (*5*)–AP (*6*)

Table 3.3 Pyrolysis of PMMA and AP in the chemical arc

	System	h, mm	u_{ox}, mm s^{-1}	u_{fuel}, mm s^{-1}	Ratio of the weight rates of linear pyrolysis	Calculated temperature of the linear pyrolysis product interaction flame, K
1	AP–PMMA (bulk)	0.5–3.0	0.30–0.15	0.17–0.08	0.34–0.32	2500–2600
2	AP–PMMA (compacted)	1.0–3.0	0.25–0.14	0.14–0.06	0.34–0.26	2500–2700
3	AP + 1 wt% CuO–PMMA (compacted)	1.0–3.0	0.45–0.32	0.17–0.10	0.23–0.19	2900
4	AP + 4 wt% CuO–PMMA (bulk)	1.5–3.0	0.55–0.52	0.14–0.12	0.15–0.14	2900
5	AP + 4 wt% Fe_2O_3–PMMA (bulk)	0.5–3.0	0.30–0.15	0.17–0.07	0.34–0.28	2500–2600
6	AP–PMMA (compacted)+ 5 wt% CuO	1.0–3.0	0.25–0.14	0.14–0.06	0.34–0.26	2500–2700
7	AP–PMMA (compacted)+ 5 wt% Fe_2O_3	1.0–2.2	0.23–0.15	0.18–0.10	0.47–0.41	2000–2300

mix flame. In this case, the linear pyrolysis of PMMA was found to occur in the third mode (see Sect. 3.2). The linear pyrolysis rate is uniquely determined by the density of the heat flux arriving at the polymer surface at almost constant temperature.

The products of PMMA linear pyrolysis undergo chemical transformations that are characteristic of fuels in standard gas flames (cracking, destruction, etc.) before they react with the products of AP gasification.

The overall AP decomposition reaction is exothermic. At high pressures, AP can burn. When heated up to 200°C or affected by a high-rate heat flux, AP can inflame and deflagrate at atmospheric pressure [15]. According to a rough estimate, the heat-flux density arriving at a linear AP pyrolysis surface from the mix flame in the chemical arc always exceeds the minimum level required for stable AP burning at atmospheric pressure. Therefore (in contrast to the fuel), the rate of AP gasification in the chemical arc is only partially affected by the heat flux from the mix flame. The decisive role is played by the heat flux generated by the deflagration flame and the heat released due to the reaction in the condensed phase of AP [6, 15, 30].

Thus, the rates of PMMA and AP burning in the chemical arc are only influenced by the addition of Fe_2O_3 and CuO if the former is introduced into the fuel and the latter into the oxidizer. Due to the hydrodynamic conditions for the flows of the gasification and burning products in the chemical arc, the catalyst is transferred to the mix flame from the fuel and oxidizer. Thus, one can assume that the changes in the PMMA and AP burning rates in the presence of Fe_2O_3 and CuO can be explained by the influence of Fe_2O_3 on processes occurring in the fuel before its involvement in the mix flame. However, this does not mean that the conditions in the mix flame are not affected by the introduction of the catalyst into the components. In the presence of Fe_2O_3 in PMMA or CuO in AP, the ratio of the weight rates of the component gasification changes. This results in a change in the component ratio in the mix flame, which leads to a change in its temperature.

The thermodynamic flame temperature for Fe_2O_3-containing PMMA was found to be approximately 500°C lower than that of the pure component burning in the chemical arc. For AP with CuO added, the thermodynamic flame temperature is 350°C higher than that of the catalyst-free system (Table 3.3).

Thus, the increase in the burning rate for the PMMA–Fe_2O_3 system takes place under the conditions of a decrease in the calculated temperature in the flame zone (Table 3.3). This can be explained by the movement of the flame zone closer to the PMMA pyrolysis surface.

Taking all of these factors into account, one can suppose that Fe_2O_3 plays the role of a catalyst for gas-phase reactions between the primary products of linear fuel pyrolysis. These catalytic processes occur before the interaction between the decomposition products of the fuel and the oxidizer and lead to the formation of more reactive components. This supposition is indirectly confirmed by a number of data. For example, introducing various additives (including Fe_2O_3) into some polymeric fuels (binders for solid propellants) was found to significantly change the chemical composition of the fuel's decomposition products [31]: the content of thermally stable compounds decreased, while the content of gaseous compounds increased.

The addition of some iron-containing agents (for example ferrocene) resulted in a marked decrease in the degree of coke formation and an increase in the yield of volatile products in diffusion flames.

In the case of the introduction of Fe_2O_3 into PMMA, the decrease in the AP burning rate can be attributed to the increase in the heat loss outward. This is caused by a reduction in the shielding effect of the interaction flame when its temperature decreases due to the change in its component ratio.

The catalytic behavior of Fe_2O_3 was studied by probing the condensed and gas phases of the AP–PMMA system with a thermocouple [32, 33, 34]. Fe_2O_3 was found to exhibit its highest catalytic activity in the near-surface layer of the condensed phase (the reaction layer of the condensed phase and the adjacent layer, $\sim 10\,\mu m$, of the gas phase).

Introduction of CuO into the oxidizer leads to an increase in the burning rate for AP and PMMA (some observations suggest that when it is added to PMMA, CuO is reduced by the products of the linear pyrolysis of PMMA before it reaches the mix flame). Thus, in contrast to Fe_2O_3, CuO promotes AP gasification and does not affect PMMA gasification under the conditions of fuel and oxidizer burning in the chemical arc. CuO was found to be an efficient catalyst for AP thermal decomposition, linear pyrolysis and burning (see Sect. **??** and [2]). In the burning of the (PMMA)–(AP+CuO) system, the increase in the PMMA gasification rate can be explained by the increase in the flame temperature. Such an increase takes place because the component ratio in the burning zone for the system with the catalyst (see Table 3.3) is more stoichiometric than that in the catalyst-free system.

Introduction of the Fe_2O_3 and CuO additives into a model solid propellant results in an increase in its burning rate [2]. However, Fe_2O_3 and CuO influence the solid propellant burning rate in different ways (as found using the chemical arc experimental method). Fe_2O_3 affects reactions between the primary products of PMMA gasification, while CuO accelerates the linear pyrolysis of AP.

The influence of Fe_2O_3 on the gas-phase kinetics is indirectly confirmed by the stronger pressure dependence of the Z_c parameter (equal to the ratio of the burning rates for a catalyst-containing system and a catalyst-free one under identical conditions [2]) for the (80% AP)–(20% PMMA+Fe_2O_3) system in comparison with the CuO-containing fuel [2]. As the pressure grows from 1 to 8 MPa, $Z_c^{Fe_2O_3}$ increases by 50% while Z_c^{CuO} increases just by 30%.

3.4 Macrokinetics of the High-Temperature Decomposition of Solid Propellant Components in Linear Pyrolysis Mode

The analysis of the experimental and theoretical data presented in Chaps. 1–3 allows one to draw some general conclusions and to compare them with literature data published after our studies of linear pyrolysis.

3.4.1 Ammonium Perchlorate

The limiting stage of AP linear pyrolysis is decomposition in the condensed phase with activation energy $E \approx 125\,\text{kJ}\,\text{mol}^{-1}$, which is in a good agreement with the values obtained in [6, 35, 36, 37]. Low (and even negative) values of the activation energy obtained in earlier works on linear pyrolysis [5, 38, 12] can be attributed to some technical aspects (neglect of the temperature drop across the gas gap, catalytic action of the heater material, and others).

When the temperature of the AP surface exceeds 500°C, the contribution of thermodynamically equilibrium dissociative sublimation (which determines the resulting degree of conversion on the material surface, η_S) becomes more significant. The content of unreacted AP in the condensed phase among the linear pyrolysis products increases. This leads to the formation of a highly calorific and combustible gas mixture (consisting of perchloric acid vapor, ammonia and products of the irreversible decomposition of AP) near the phase interface. For AP burning at pressures of 0.05–0.5 MPa, the surface temperature of AP is mainly determined by the gasification of some of the compound as it undergoes equilibrium dissociation [39, 40]. However, in this case, the burning rate depends on both the kinetics of the gas-phase reactions and the kinetics of the processes in the condensed phase accompanied by dispersion [15]. During linear pyrolysis, as well as when burning occurs, very high degrees of material decomposition can be attained (in the limiting case, $\eta_S \to 1$). Because of this, one can use data on the temperature dependence of the decomposition rate constant obtained by linear pyrolysis in order to analyze the mechanism of composite solid propellant burning (for example, to estimate the degree of conversion of the oxidizer in the reaction layer of a composite solid propellant). On the other hand, some additional data on the formal kinetics of high-temperature decomposition, $\varphi(\eta)$, are required to determine the actual value of η_S(which would for example clarify the contribution of dissociative sublimation to the overall process). These data cannot be obtained by linear pyrolysis, which represents a "discrete" method [41]. Some nonisothermal kinetic techniques can provide the required information about the formal kinetics and the effect of pressure on the macrokinetics of the high-temperature decomposition of AP (see Chap. 8).

3.4.2 Polymethylmethacrylate

According to the linear pyrolysis theory, the linear decomposition of PMMA (in qualitative agreement with theory) occurs in one out of the three regimes: the inner kinetic mode (dominated by outward heat losses), the inner diffusion mode (no heat losses) and the so-called third mode (similar to the second one, but with significantly higher kinetic constants E and k_0). Since the linear pyrolysis rates are quite high in a solid-propellant rocket engine, and the influence of the outward heat loss from the decomposing fuel can be neglected, the second and especially third decomposition modes are of practical interest (studying the linear pyrolysis by the method described in Sect. 1.5 using installation LP-4, see Fig. 2.7, is of special interest from a pure kinetics viewpoint). At $T_S \geq 800$–$850\,K$, a change in the kinetic mechanism of PMMA (linear and crosslinked) decomposition accompanied by an increase in k_0 from 10^{12}–$10^{13}\,s^{-1}$ to 10^{21}–$10^{23}\,s^{-1}$ and higher and in E from 180 to 290–330 $kJ\,mol^{-1}$was found to take place. Therefore, the extrapolation of data obtained by standard kinetic methods in the low-temperature region (used for the analysis of polymer burning) should be performed very cautiously. There is every reason to believe that at low temperatures the relative content of the polymer that decomposes with such high k_0 and E is negligibly small. However, as T_S increases, the content of this polymer in the system grows and approaches 100%.

The third mode is characterized by a high activation energy, close to the energy of main-chain C–C bond rupture. This allows one to assume that the third mode of PMMA linear pyrolysis (and the corresponding kinetic constants) should be characteristic of the linear pyrolysis of other polymers used as fuels in composite solid and hybrid propellants. From a kinetics viewpoint, this supposition is related to the change in the mechanism of polymer molecule decomposition when random bond opening becomes the dominant stage in the chain process. The experimental results, which show a weak pressure dependence of the rate of polymer linear pyrolysis, agree with this idea. These conclusions [42] allow one to explain the following general patterns that are characteristic of a number of polymers that differ significantly in terms of their structures and constants of low-temperature decomposition:

a) Comparatively low T_S (measured by placing microthermocouples into the burning polymer) at linear burning rates $U \approx 10^{-3}$–$10^{-2}\,m\,s^{-1}$ [26, 43]. Here $T_S = 560°C$ when calculated from experimental data on the pulsed linear pyrolysis of PMMA at $U \approx 0.1$–$0.2\,m\,s^{-1}$ in the chamber of a pulsed plasma electric propulsion [44, 45].
b) The so-called Sreznevsky law for burning spherical polymer particles [46]. This law (a linear reduction in the spheres' surface area as a function of time) is valid only for insignificant variations in T_S over a wide range of the pyrolysis rate which increases as

$$U \sim (d_0^2 - \kappa t)^{-1},$$

where d_0 is the initial sphere diameter, t is the time, and κ is the Sreznevsky constant.
c) Insignificant influence of P_∞ on the rate of hybrid propellant burning, which can be explained purely by heat exchange factors.

The ideas about the high-temperature decomposition of polymers formulated above were later clearly confirmed in [28, 47, 48, 49, 50, 51, 52, 53]. In these works, the number of studied polymer classes was significantly expanded. Also products of their decomposition were analyzed. The decomposition of polymers was investigated by isothermal methods over a wide range of pressures and for temperatures of up to 500°C (for PMMA). The decomposition rate was found to be almost independent of the pressure in the range of $P_\infty = 0.03$–$5\,MPa$ for PMMA, synthetic rubber (divinyl and isoprene) and polyethylene [52]. The change in the main decomposition stage for the mentioned polymers was demonstrated by an increase in the activation energy and by a dramatic change in the gaseous product composition. In our opinion, this is the most significant result. A summary of the data on the constants characterizing various independent parallel reactions of high-temperature polymer decomposition, as obtained by Reshetnikov [27], are presented in Table 3.4. These data are in good agreement with the data obtained for PMMA by the linear pyrolysis method. It is interesting that the kinetic constants for the second reaction are quite similar for all polymers, even those that are significantly different in nature ($k_{02} \approx 10^{23}$–$10^{24}\,\mathrm{s}^{-1}$, $E_2 \approx 330$–$350\,\mathrm{kJ\,mol}^{-1}$). The kinetic constants for the first (low-temperature) reaction are drastically different ($k_{01} \approx 10^{13}$–$10^{19}\,\mathrm{s}^{-1}$, $E_1 \approx 175$–$280\,\mathrm{kJ\,mol}^{-1}$). The difference between the $k_1(T)$ values at similar values of $k_2(T)$ determines the difference between the temperatures at which the mechanism changes (T_{ch}), which can be considered the temperatures at which the third mode of linear pyrolysis begins (T_S^{III}). The corresponding values of T_{ch} estimated from the data in [27] using the obvious relationship $k_1(T_{\mathrm{ch}}) = k_2(T_{\mathrm{ch}})$ are shown in Table 3.4. $T_{\mathrm{ch}}^{\mathrm{PMMa}} = 550°\mathrm{C}$ is in good agreement with the value obtained above (520°C, see Sect. 3.2.2). Thus, the decomposition macrokinetics for polymeric fuels (used in hybrid propellants) are completely determined by the rate of heat transfer from the flame and by the k_{02} and E_2. The macrokinetics of the high-temperature decomposition of fuels used as binders in composite solid propellants are completed by reactions in the condensed phase such as the decomposition of the oxidizer and heterogeneous processes in the fuel–oxidizer system (see Chaps. 5 and 9 and [54]).

If the third mode characterized by $T_{\mathrm{S}}^{\mathrm{III}} = T_{\mathrm{ch}}$ is assumed to be a special feature of linear pyrolysis at polymer burning, an important conclusion regarding polymer burning in the combustion chambers of hybrid rocket engines can be made. In this case, the burning rate should weakly depend on the kinetics of high-temperature polymer decomposition and be mainly determined by the thermophysical and thermodynamic parameters that dictate the heat balance for linear pyrolysis. The general heat-balance equation for the surface of the burning hybrid propellant block is

$$q_{\mathrm{gph}} = \dot{m} Q_{\Sigma},$$

where q_{gph} is the density of the heat flux from the gas phase onto the linear pyrolysis surface, $\dot{m}$ is the weight rate of the linear pyrolysis (burning), and Q_{Σ} is the total heat of polymer gasification in the linear pyrolysis mode.

Table 3.4 Kinetic constants of the high-temperature decomposition of polymers [27]

	Polymer	k_{01}, s^{-1}	E_1, $\mathrm{kJ\,mol}^{-1}$	k_{02}, s^{-1}	E_2, $\mathrm{kJ\,mol}^{-1}$	$T_{\mathrm{ch}} = T_{\mathrm{S}}^{\mathrm{III}}$
1	PMMA	1.2×10^{13}	178	1.1×10^{23}	335	550
2	Synthetic divinyl rubber	1.0×10^{16}	260	4.5×10^{23}	360	400
3	Synthetic isoprene rubber	1.4×10^{15}	220	10^{24}	355	515
4	Polyethylene	2.2×10^{19}	280	2.5×10^{23}	350	610

The unimportant role of the kinetic factors is explained by the fact that they only influence T_S^{III}, which contributes insignificantly to the total gasification heat $Q_\Sigma = \Delta H + c_s(T_S^{III} - T_\infty)$, where ΔH is the thermal effect of high-temperature decomposition and $c_s(T_S^{III} - T_\infty)$ is the heat of warming-up.

The degree of reliability of this conclusion is of special interest when attempting to understand the mechanism of polymer burning and when developing methods for increasing the rate of burning in hybrid rocket engines.

The value of Q_Σ cannot be theoretically estimated at present due to a lack of data on high-temperature linear pyrolysis product composition for most practically important polymeric fuels. The next chapter is devoted to methods for the experimental determination of Q_Σ as well as the verification of the ideas discussed in this chapter.

References

1. Beckstead MW (2000) In: 29th JANNAF Propellant Development & Characterisation Meeting, Cocoa Beach, FL, USA, 8–12 May 2000
2. Bakhman NN, Belyaev AF (1967) Combustion of heterogeneous condensed systems. Nauka, Moscow
3. Korobeynichev OP (1971) (ed) Mechanism, kinetics and catalysis of thermal decomposition and combustion of ammonium perchlorate. Nauka, Novosibirsk
4. Schultz RD, Dekker AO (1955) In: Proceedings of the 5th Symposium on Combustion. Reinhold, New York
5. Chaiken RF (1959) Combust Flame 3:285
6. Manelis GB, Nazin GM, Rubtsov YI, Strunin VA (2003) Thermal decomposition and combustion of explosives and propellants. Taylor & Francis, New York
7. Shaw RW, Brill TB, Thompson DL (2005) (eds) Advanced series in energetic materials: overviews of recent research on energetic materials, vol 16. World Scientific, Singapore
8. Shteinberg AS (1980) In: Merzhanov AG (ed) Heat and mass exchange in combustion processes. Nauka, Chernogolovka, p 85
9. Manelis GB, Strunin VA, Shteinberg AS (1968) Combust Explos Shock Waves 4:305
10. Bills KW, Mishuck E, Moe G, Schultz RD (1959) Combust Flame 3:301
11. Ishihara A, Sarai Y, Konishi K, Andoh E (2005) Proc Combust Inst 30:2123
12. Guinet M (1955) Fr Rech Aerospat 10:41
13. Jnami S, Rosser W, Weis NM (1963) J Phys Chem 67:1077
14. Manelis GB, Strunin VA (1971) Combust Flame 17:69
15. Maksimov EI, Grigor'ev YuM, Merzhanov AG (1966) Izv Akad Nauk SSSR Ser Khim 3:422
16. Shteinberg AS (1979) DSci thesis. State Institute of Applied Chemistry, Institute of Chemical Physics of the USSR Academy of Sciences, Chernogolovka, Russia
17. Grassi N (1959) Chemistry of polymer destruction. Inostrannaya Literatura, Moscow
18. Madorsky SL (1964) Thermal degradation of organic polymers. Wiley, New York
19. Blatz PJ (1963) 6th AGARD Colloquium. Pergamon, London
20. Zenin AA (1976) DSci thesis. Institute of Chemical Physics of the USSR Academy of Sciences, Moscow
21. Committee on Advanced Energetic Materials and Manufacturing Technologies (2004) Advanced energetic materials. Committee on Advanced Energetic Materials and Manufacturing Technologies, National Research Council of the National Academies, Washington, DC
22. Davenas A (1993) Solid propellant propulsion technology. Pergamon, Oxford
23. Chaiken RF (1962) J Chem Phys 37:2311

24. Cantrell RH (1963) AIAAJ 1:1544
25. Tkachenko EV, Ulybin VB, Shteinberg AS (1969) Combust Explos Shock Waves 5:1116
26. Shteinberg AS, Ulybin VB, Manelis GB, Dolgov EI (1972) In: Stesik LN (ed) Combustion and explosion. Nauka, Moscow, p 124
27. Reshetnikov SM, Morozova ZG (1979) Seminar on combustion of polymers and development of restrictedly combustible materials. Volgograg, Russia
28. Reshetnikov SM (1975) PhD thesis. Kazan' Institute of Aviation, Kazan'
29. Shteinberg AS (1967) PhD thesis. Institute of Chemical Physics of the USSR Academy of Sciences, Chernogolovka, Russia
30. Manelis GB, Strunin VA (1972) In: Stesik LN (ed) Combustion and explosion. Nauka, Moscow, p 53
31. Rubtsov YuI, Manelis GB (1964) Rus J Phys Chem 37:1292
32. Puchkov VM (1978) PhD thesis. Institute of Chemical Physics of URRS Academy of Sciences, Chernogolovka, Russia
33. Zenin AA, Leypunsky OI, Puchkov VM (1971) Dokl Phys Chem 198:400
34. Puchkov VM, Zenin AA, Leypunsky OI (1972) In: Stesik LN (ed) Combustion and explosion. Nauka, Moscow, p 74
35. Manelis GB, Rubtsov YuI (1966) Rus J Phys Chem 40:416
36. Jacobs PW, Russel-Jones A (1967) AIAAJ 5:829
37. Jacobs PW, Russel-Jones A (1968) J Phys Chem 72:202
38. Barrer M, Liberer M (1965) In: 25th Meeting of the AGARD Combustion Panel, La Jolla, USA
39. Powling J, Smith WA (1962) Combust Flame 6:173
40. Powling J, Smith WA (1963) Combust Flame 7:269
41. Merzhanov AG (1973) Combust Explos Shock Waves 9:4
42. Shteinberg AS, Ulybin VB (1969) In: 2nd All-Union Symp on Combustion and Explosion, Chernogolovka, Russia
43. Alekseev YuI, Korolyov VL, Knyazhitsky VP (1972) In: Stesik LN (ed) Combustion and explosion. Nauka, Moscow, p 128
44. Rusanov VP, Popov NP (1971) Teplofiz Vys Temp 1:9
45. Rusanov VP, Popov NP (1971) Teplofiz Vys Temp 2:40
46. Essenhigh RH, Lance Dreier W (1967) 5th Aerospace Meeting, New York, USA
47. Shchukin VK, Reshetnikov SM (1973) Trans Kazan' Inst Aviat 154
48. Reshetnikov SM (1975) Trans Kazan' Inst Aviat 184
49. Kirsanov YuA, Shchukin VK, Reshetnikov SM (1975) Combust Explos Shock Waves 10:15
50. Shchukin VK, Dresvyannikov FN, Kirsanov YuA (1973) College Trans Aviat Eng 4
51. Kirsanov YuA (1973) Trans Kazan' Inst Aviat 158
52. Kirsanov YuA (1973) PhD thesis. Kazan' Institute of Aviation, Kazan'
53. Shlensky OF, Aksenov LN, Shashkov AG (1991) Thermal Decomposition of Materials. Effect of Highly Intensive Heating. Elsevier, Amsterdam
54. Braginsky VI, Bakhman NN (1971) Combust Explos Shock Waves 7:240

Chapter 4
Study of the Total Heat of the High-Temperature Linear Pyrolysis of Polymeric Materials as Applied to Burning in Hybrid Rocket Engines

Abstract An original block calorimeter for measuring the total heat from the high-temperature linear pyrolysis of polymeric materials is described. This value includes the thermal effect of fast high-temperature decomposition and the heat required to preheat the polymer up to an almost constant pyrolysis surface temperature. The latter is characteristic of the "third mode" of linear pyrolysis in which fast burning-induced destruction of almost all polymers occurs. The theoretical background for the application of the instrument and the method used to calibrate it are presented. Data on the total heat of high-temperature linear pyrolysis for more than 20 polymeric materials are given. Data on the combustion of numerous polymer fuels in a model hybrid rocket engine (the most promising environmentally safe rocket engine) are presented. The value of the total heat of high-temperature linear pyrolysis is shown to be the main parameter that determines the rate of polymer burning in the hybrid rocket engine. Thus data on the total heat can be used to develop polymer fuels characterized by optimal rates of burning in the hybrid rocket engine.

4.1 Block-Calorimeter for Measuring the Total Heat of the High-Temperature Linear Pyrolysis of Polymers

A special block calorimeter (Fig. 4.1) was used to measure the total heat values, Q_Σ [1]. The principle of operation of the device is based on the measurement of the change in the total heat of a hot block used as a heater for the material under study (linear pyrolysis of the sample occurs at one of the block ends). The calorimeter consists of a bulky metal cylinder (40 mm in diameter, 180 mm long) *6* placed into a quartz Dewar vessel *5*. The temperature on the external surface of the Dewar vessel is kept constant using a simple dual control system. The hot ends of working thermocouples are calked into the top, middle and bottom sections of the block. A sample is pressed onto a top end of the hot block (in experiments with softening polymers, a block with a small sample-containing cavity on the top end is used). The cooling of the block that provides the heat for linear pyrolysis is recorded by a multichannel recorder (EPP-09). Q_Σ can be calculated from the heat change of the block and the weight of the decomposed polymer sample, m_0 (determined as

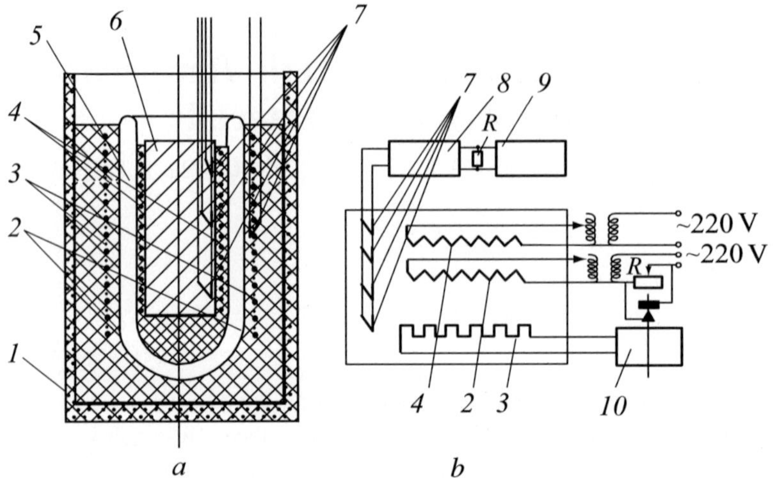

Fig. 4.1 Schematic of the high-temperature block calorimeter: asbestos cement casing (*1*), external heater (*2*); resistance thermometers (*3*); inner heater (*4*), quartz Dewar vessel (*5*), metal cylinder (duralumin or copper) (*6*), thermocouples (*7*), photoamplifier (*8*), recorder (*9*), control relay (*10*)

the difference between the values of the sample weight before and after the experiment). An additional internal heater on the lateral surface of the block is used for calorimeter standardization (determining the heat exchange parameters in the Dewar vessel at high temperatures and measuring the effective heat capacity of the block calorimeter). The expression for calculating Q_Σ from the experimental data was found by solving the problem of nonstationary heat conduction in the block.

Two process stages were considered: linear pyrolysis and the subsequent thermal relaxation. The solution obtained for the first stage was used as the initial condition for the solution of the problem related to the second stage. In order to simplify the resulting analytical expression the following assumptions were made (detailed information regarding the analytical method can be found in [2]):

a) The model used is one-dimensional (any cross-sectional temperature difference is much lower than that longitudinally)
b) The linear pyrolysis rate is constant
c) The rates of heat dissipation from the block top end before and after linear pyrolysis are equal

The assumptions are quite reasonable due to a big ($\sim$ 25-fold) difference in the corresponding Fourier numbers (relates to assumption a) and to an insignificant variation in the block end temperature during linear pyrolysis ($\sim$ 10–15°C) (relates to assumptions b, c). Typical time dependencies for the block temperature during the linear pyrolysis of PMMA are shown in Fig. 4.2. The maximum temperature change in the top section of the block is equal to 11°C (from 493 to 482°C). Shortly after linear pyrolysis, the thermal relaxation of the block was found by numerical analysis

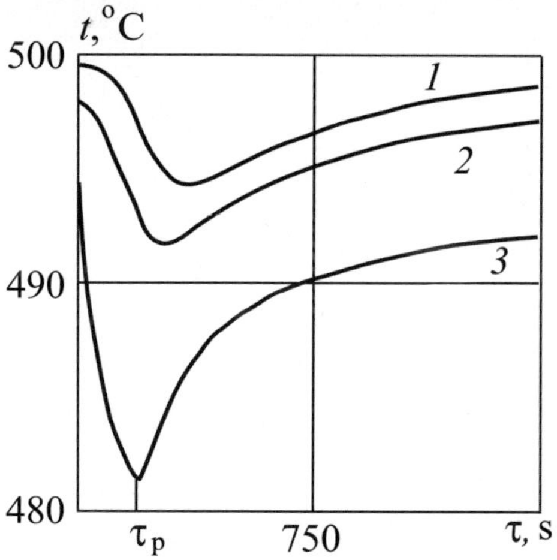

Fig. 4.2 The temperature change in the block during the linear pyrolysis of PMMA ($m_0 = 3.93$ g) and at thermal relaxation for the bottom section (*1*), the middle section (*2*), and the top section (*3*)

to occur in the so-called *regular* mode. In this case, the average temperature in the block longitudinal section is close to its volumetric mean value.

The expression for Q_Σ calculation is

$$Q_\Sigma = \frac{(\Sigma MC)_{\text{cal}} \Delta T_0}{m_0 \mu \varepsilon}, \tag{4.1}$$

where $(\Sigma MC)_{\text{cal}}$ is the effective heat capacity of the calorimeter, ΔT_0 is the calculated (see below) change in the middle ($l/2$) block section, $\mu = \cos(\nu_0/2)$ is the correction for the heat transfer through the Dewar vessel bottom, $\varepsilon = \frac{\exp\left[(\text{Bi}+\nu_0^2/\beta^2)\tau_{\text{p}}\right]-1}{(\text{Bi}+\nu_0^2/\beta^2)\tau_{\text{p}}}$ is the correction for the heat transfer through the Dewar vessel walls during pyrolysis [$\text{Bi} = \alpha\delta/\lambda$ is the Bio number, $4\delta = d$ is the block diameter, $\beta = l/\delta$ is the geometrical block parameter, $\tau_{\text{p}} = at_{\text{p}}/\delta^2$ is the dimensionless time (Fourier number) of linear pyrolysis, ν_0 is the minimum eigenvalue for $\nu_0 = \beta\text{Bi}/\text{tg}\nu_0$, α is the factor for heat transfer through the Dewar vessel walls, λ and a are the thermal conductivity and the temperature diffusivity of the block material, and t_{p} is the duration of linear pyrolysis].

ΔT_0 is determined from the time dependence of $\lg\Delta T$ (Fig. 4.3) by extrapolating the straight line corresponding to a "regular" mode to $t = 0$. In the case of $\varepsilon \approx 1$ (fast pyrolysis, good thermal insulation of the block), the expression

$$Q_\Sigma = \frac{(\Sigma MC)_{\text{cal}} \Delta T_{0.5}}{m_0 \mu}, \tag{4.2}$$

where $\Delta T_{0.5}$ is determined by the same extrapolation method to $t = 0.5t_{\text{p}}$, can be used instead of Eq. (4.1).

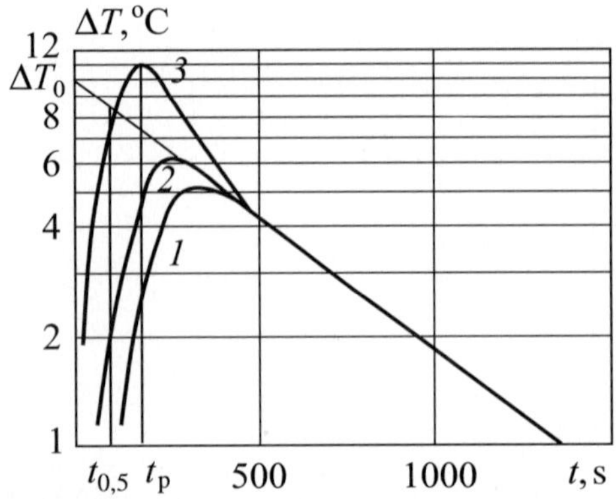

Fig. 4.3 Determination of ΔT_0: analysis of data shown in Fig. 4.2 for the bottom section (*1*), the middle section (2), and the top section (*3*), $\Delta T_0 = 9.71°\mathrm{C}$

The block calorimeter was standardized by the method of pulsed switching-on of the internal heater (the measured constants are shown in Table 4.1). The temperature dependence of $(\Sigma MC)_{\mathrm{cal}}$ is in good agreement with literature data on the temperature dependence of the specific heat capacity ($C = C(T)$) for duralumin and copper. The increase in the Bio number with increasing temperature is explained by the growth of the irradiation contribution to the overall heat transfer through the Dewar vessel walls.

Table 4.1 Block-calorimeter constants determined by the method of pulsed heating followed by relaxation

Material of the block	Temperature, °C	$(\Sigma MC)_{\mathrm{cal}}$, J(°C)$^{-1}$	Bi
Duralumin	130	621	3.2×10^{-4}
	485	750	1.1×10^{-3}
Copper	400	1854	8.3×10^{-4}
	685	2125	1.5×10^{-3}

In Tables 4.2 and 4.3, Q_Σ values for a number of model polymer systems, commercial hybrid propellants and some organic compounds are given.

4.2 Correlation Between the Burning Rates of Polymer Hybrid Propellants and the Total Heats of Linear Pyrolysis

One of the most interesting aspects of polymer burning is the relationship between its burning rate and its decomposition characteristics. The burning rates for some materials have been measured (under the standardized conditions of the combustion chamber of a model hybrid rocket engine) to check the conclusion about the dominant role of the total heat of linear pyrolysis in the determination of the burning

Table 4.2 Total heats of linear pyrolysis for hybrid propellant polymers and fuels

Material	Material of the block	T_0, °C	Q_Σ, J g^{-1}
PMMA (linear)	Duralumin	485	1583 ± 40
PMMA (linear)	Copper	665	1666 ± 40
PMMA + 2 wt % triethyleneglycoldimethacrylate	Copper	665	1480 ± 20
PMMA + 10 wt % triethyleneglycoldimethacrylate	Copper	665	1540 ± 20
Polyethylene	Copper	665	2354 ± 40
Polystyrene	Copper	665	1875 ± 40
Polyethyleneimine	Duralumin	485	1646 ± 40
Rubber resin–SKDN	Duralumin	485	625 ± 20
Rubber resin–PDI-1	Duralumin	485	875 ± 20
Diene polymer	Duralumin	485	1041 ± 40
30 wt% rubber resin–SKDN + 70 wt% transformer oil	Duralumin	485	980 ± 20
40 wt% rubber resin–SKDN + 60 wt% transformer oil	Duralumin	485	625 ± 20
60 wt% rubber resin–SKDN + 40 wt% transformer oil	Duralumin	485	646 ± 20
Transformer oil	Duralumin	350	1333 ± 40
Polyurethane–DT	Duralumin	485	1541 ± 40
Chloropolyurethane	Duralumin	485	875 ± 40
20 wt% organic azide–APU + 80 wt% polyurethane–DT	Duralumin	485	790 ± 20
64 wt% organic azide–APU + 16 wt% rubber resin–PDI-1 + 20 wt% polyethylene	Duralumin	485	416 ± 20
20 wt% organic azide–APU + 80 wt% rubber resin–SKDN	Duralumin	485	333 ± 20

rate. In the experiments, cylindrical samples (100 mm long, initial hole diameter of 13 mm) were burned in flowing oxygen at an O_2 pressure of 1.5 MPa and a flux density of 60 g cm^{-2} s^{-1}.

A plot of the experimental weight-average burning rate ($\dot{m} = \bar{u}\gamma_s$, where $\bar{u}$ is the average burning rate for the polymer fuel and γ_s is the density of the polymer fuel) against the corresponding total heat of linear pyrolysis is shown in Fig. 4.4.

The complicated mechanism of blowing-assisted burning in the interface layer that is characteristic of a hybrid rocket engine is not considered in detail here. However, one can draw a conclusion about the decrease in the burning rate of a polymer fuel with increasing Q_Σ. The correctness of this conclusion is confirmed by the fact

Table 4.3 Total heats of linear pyrolysis for some organic compounds

Compound	Block material	T_0, °C	Q_Σ, J g^{-1}
Benzyl	Duralumin	400	1000 ± 40
Diphenyl	Duralumin	400	958 ± 20
Sebacic acid	Duralumin	400	1646 ± 62
Stilbene	Duralumin	400	1000 ± 40
Diphenylamine	Duralumin	400	980 ± 20
Succinic acid	Duralumin	400	1416 ± 20
Glutaric acid	Duralumin	400	1354 ± 20
Adipinic acid	Duralumin	400	1416 ± 20
Azobenzene	Copper	665	1000 ± 40
Cinnamic acid	Copper	665	1500 ± 83
Fumaric acid	Copper	665	1375 ± 40
Para-phenylenediamine	Copper	665	1290 ± 62

that the experimental data in Fig. 4.4 correspond to systems that differ greatly in terms of their structure and low-temperature decomposition constants.

Since this monograph is mostly dedicated to the macrokinetics of high-temperature decomposition in the condensed phase, gas-phase processes in a hybrid rocket engine are not analyzed (additional information on this topic can be found in [3, 4, 5, 6]). However, the result obtained is quite important for the theory of combustion in a hybrid rocket engine and for developing methods of increasing the hybrid propellant burning rate. The decrease in Q_Σ due to, for example, the addition of an organic compound that decomposes and releases heat into the fuel results in a

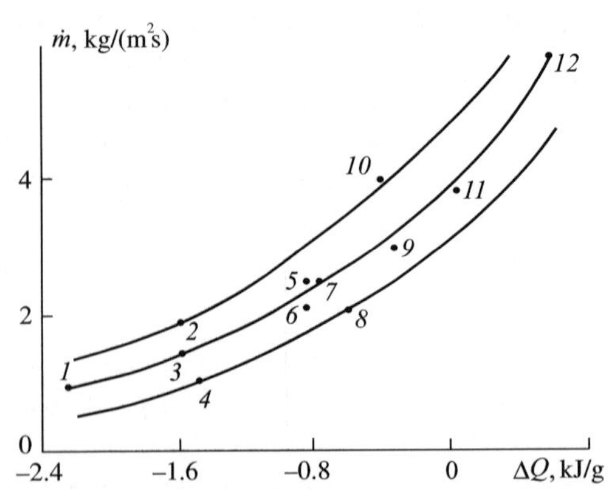

Fig. 4.4 Correlation between the rates of burning in a hybrid rocket engine and the total heats of linear pyrolysis for polymeric fuels at the oxidized flux density $G_{ox} = 60\,\mathrm{g\,cm^{-2}\,s^{-1}}$ and $P = 1.5\,\mathrm{MPa}$ for polyethylene (*1*), polyethyleneimine (2), PMMA (*3*), polyurethane–DT (*4*), chloropolyurethane (*5*), rubber resin PDI-1 (*6*), 20% organic azide–APU + 80% polyurethane–DT (*7*), rubber resin–SKDN (*8*), 20% organic azide–APU + 80% rubber resin–SKDN (*9*), 64% organic azide–APU + 16% rubber resin–PDI-1 + 20% polyethylene (*10*), 40% organic azide–APU + 60% rubber resin–SKDN (*11*), 80% organic azide–APU + 20% rubber resin–PDI-1 (*12*)

significant increase in the rate of linear pyrolysis in a hybrid rocket engine. This can be exemplified by the increase in the linear pyrolysis rate for rubber resin–PDI from 2 to 4 mm s^{-1} (under quite moderate conditions in the chamber) due to the addition of organic azide–APU (40 wt%) into the fuel. In this case, the oxidizer shutdown does not cause a change in the process mode to self-sustaining gas generation.

The universal character of the $\bar{u}\gamma(Q_\Sigma)$ dependence serves as the basis for an express method for the preliminary evaluation of the hybrid propellant burning rate from the experimental data on the total heat change during high-temperature linear pyrolysis using a block calorimeter designed by us.

No correlation between Q_Σ (Tables 4.2 and 4.3) and burning rates was found for composite solid propellants based on the corresponding fuel-binders and AP [7]. This agrees with modern ideas about the dominant role of the macrokinetics of reactions in the condensed phase and gas-phase interactions between linear pyrolysis products in determining the burning patterns for composite solid propellants. Some techniques for studying fast exothermal decomposition kinetics for model and industrial energetic materials are considered in Chaps. 5, 8 and 9.

References

1. Dmitriev BM, Kochetov OA, Ulybin VB, Shteinberg AS (1969) Combust Explos Shock Waves 5:17
2. Carslaw HS, Jager JC (1959) Conduction of heat in solids, 2nd edn. Clarendon, Oxford
3. Shteinberg AS, Ulybin VB, Manelis GB, Dolgov EI (1972) In: Stesik LN (ed) Combustion and explosion. Nauka, Moscow, p 124
4. Alekseev YuI, Korolyov VL, Knyazhitsky VP (1972) In: Stesik LN (ed) Combustion and explosion. Nauka, Moscow, p 128
5. Marxman GA, Muzzy RL (1967) ICRPG/AIAA 2nd Solid Propulsion Conf, Anaheim, CA, USA, 6–8 June 1967
6. Kustov YuA, Rybanin SS (1970) Combust Explos Shock Waves 6:50
7. Braginsky VI, Bakhman NN (1971) Combust Explos Shock Waves 7:240

Chapter 5
Study of the Initial-Stage Kinetics for High-Temperature Decomposition of Homogeneous Blasting Powders and Composite Solid Propellants by the Method of Ignition Using a Heat-Conducting Block

Abstract An ignition theory for an energetic material sample at a given constant instantaneous surface temperature is presented. An expression for calculating the rate constant of the fast high-temperature decomposition of the energetic material from experimental data on the ignition delay period as a function of the surface temperature is given. An original flash-block instrument (in which a sample is pressed onto the end of a thermostated bulky Al cylinder) is described. Data on the ignition and kinetics of the fast high-temperature decomposition of some homogeneous and heterogeneous energetic materials including nitrocellulose and model and commercial solid propellants are given. The activation energies and pre-exponential factors of most high-temperature decomposition reactions were shown to differ significantly from those occurring in the low-temperature region.

5.1 Application of Ignition Methods to Studies of the Kinetics of Condensed Material High-Temperature Decomposition

The applicability of isothermal kinetic methods to studies of fast high-temperature exothermal decomposition of energetic materials is limited due to the ignition and thermal explosion phenomena that are characteristic of such systems. On the other hand, one can obtain information on effective formal kinetic constants, provided that the thermal explosion and ignition parameters are clearly defined in preset thermophysical conditions.

Methods for determining kinetic characteristics from the critical ignition parameters of a material layer in which the interfaces are kept at (different) constant temperatures are based on a classical work by Zel'dovich [1]. The results from experimental and theoretical studies in this area were reviewed in [2, 3, 4].

Here data on ignition in the combustion chamber of a solid-propellant rocket engine and on the ignitions of powders with hot gas flow, hot wires, radiation heating and shock waves [4] are not considered. In these works, the ignition is assumed to be of a thermal nature because the thermal decomposition is characterized by a low degree of conversion in the volume of the thin material layer near the heated material surface.

The thermal ignition theory [5] allows one to find a relationship between the kinetic constants of the limiting reaction, the thermophysical parameters of the material, the heat-exchange conditions on its surface and the ignition delay t_{ign}. According to the thermal ignition theory [4], the temperature field in the material can be described by a nonstationary thermal conductivity equation for an inert body up to the moment of ignition (up to the inflection point of the dependence of the surface temperature on time). Therefore, the parametric relationship between the macro characteristics of the ignition and the kinetic constants of the reaction strongly depends on the behavior of the heat flux as a function of time. Under the conditions of a decreasing heat-flux density, t_{ign} is strongly influenced by the kinetic parameters (for constant thermophysical characteristics of the material). In contrast, at increasing heat-flux density, the kinetic constants of high-temperature decomposition only slightly affect t_{ign} [6, 7] (this situation is characteristic of ignition in the combustion chamber of a solid-propellant rocket engine). In the latter case, information on the process kinetics cannot be obtained from the data from ignition experiments.

These circumstances were taken into consideration when developing a method for studying the fast exothermal reaction kinetics of condensed materials based on their ignition parameters at a constant material surface temperature [8, 9]. The material surface was instantly heated up to a high temperature (which was kept constant) by pressing the sample onto the end of a bulky heat-conducting block (similar to a heating block in installation LP-4 described in Sect. 2.2). Strictly speaking, in this case, boundary conditions of the fourth kind are valid for the surface of the heated material (rather than the boundary conditions of the first kind). However, due to the huge difference between the thermal activities of aluminum (the block material) and the ignited materials (homogeneous and composite powders), thermostating via the contact (boundary conditions of the fourth kind) provides conditions close to ideal heat exchange with $\mathrm{Bi} \to \infty$ (boundary conditions of the first kind). Indeed, according to the theory of heat conduction [8], the temperature drop (ΔT) at the surface of a hot body after its contact with a cold (T_∞, j_2) semi-infinite continuum is

$$\Delta T = \frac{T_0 - T_\infty}{1 + \sqrt{j_1/j_2}}, \tag{5.1}$$

where

$$j = \lambda c \rho \tag{5.2}$$

is the thermal activity, T_0 is the hot body initial temperature, and j_1 is the hot body thermal activity.

Substituting in corresponding values for aluminum [10] ($\lambda_1 = 234\,\mathrm{W\,m^{-1}\,{}^\circ C^{-1}}$, $c_1 = 1.04\,\mathrm{kJ\,kg^{-1}\,{}^\circ C^{-1}}$, $\rho_1 = 2.7 \times 10^3\,\mathrm{kg\,m^{-3}}$) and for a typical polyurethane composite solid propellant (NG-2-60-DTM) determined experimentally ($\lambda_2 = 0.326\,\mathrm{W\,m^{-1}\,{}^\circ C^{-1}}$, $c_2 = 1.30\,\mathrm{kJ\,kg^{-1}\,{}^\circ C^{-1}}$, $\rho_2 = 1.76 \times 10^3\,\mathrm{kg\,m^{-3}}$), one obtains $\Delta T_{\mathrm{block}}^{\mathrm{calc}} = 12^\circ\mathrm{C}$ for $T_0 = 400^\circ\mathrm{C}$ and $T_\infty = 20^\circ\mathrm{C}$. The experimentally determined temperature drop ($\Delta T_{\mathrm{block}}^{\mathrm{exp}} = 2 - 3^\circ\mathrm{C}$) is lower due to intensive heat pumping from the periphery to the sample–block contact (the area of the end of the block is 16 times larger than that of the sample–block contact). This systematic inaccuracy does not cause any significant errors when estimating the kinetic parameters.

When analyzing experimental data, one should also evaluate the limits of the applicability of the homogeneous model for ignition in real heterogeneous composite solid propellants at $T_0 = \text{const}$. The model is believed to be valid if the reaction layer thickness exceeds the maximum characteristic size of the AP crystal. For most of the ignition delay period, the temperature field in the composite solid propellant can be described using

$$\frac{T - T_\infty}{T_0 - T_\infty} = erf\left(\frac{x}{2\sqrt{a_s t_{ign}}}\right), \tag{5.3}$$

where a_s is the temperature diffusivity of the sample.

The upper limit of t_{ign} is important for homogeneous systems, since it determines the applicability of the semi-infinite continuum preheating and ignition model to samples of a finite length l_0 (for composite solid propellants and powders at $l_0 = 1$ cm the upper limit of t_{ign} is in the range of 150–200 s). Assuming that the temperature change in the reaction layer is equal to one Semenov interval RT_0^2/E (with an accuracy that is acceptable for approximate estimation), one obtains for the reaction layer thickness at the moment of ignition, Δ_{ign},

$$\frac{RT_0^2}{E(T_0 - T_\infty)} = \Phi\left(\frac{\Delta_{ign}}{2\sqrt{a_s t_{ign}}}\right). \tag{5.4}$$

For standard composite solid propellants with high contents of fine-grained AP, one can suppose that $\Delta_{ign} \geq 10^{-4}$ cm. Thus, for example, for a typical polyurethane composite solid propellant (with the thermophysical characteristics given above and $E \approx 165\,\text{kJ mol}^{-1}$), the homogeneous model is valid for $t_{ign} \geq 3.5$ s.

An expression obtained by Averson, Barzykin and Merzhanov [11] (who studied ignition at $T_0 = \text{const}$ for a first-order reaction) in order to analyze the experimental data is

$$\tau_{ign} = (1 + 1.6\Theta_{0\infty} + 0.2\Theta_{0\infty}^2)(1 + 8\Theta_{0\infty}\gamma^{3/2})(1 + \beta), \tag{5.5}$$

where

$$\Theta_{0\infty} = \frac{E}{RT_0^2}(T_0 - T_\infty); \gamma = \frac{c}{Q}\frac{RT_0^2}{E}; \beta = \frac{RT_0}{E}; \tau_{ign} = t_{ign}\frac{Qk_0E}{cRT_0^2}\exp\left(-\frac{E}{RT_0}\right).$$

Equation (5.5) was obtained by the approximation of a numerical solution for a complete set of thermal conductivity and kinetic equations. An analytical expression for the $\varphi(\eta)$ dependence cannot be obtained by the ignition methods. However, for a wide range of non-self-catalyzed reactions occurring at $T = \text{const}$, the inaccuracy caused by the a priori assignment of $\varphi(\eta)$ does not result in serious errors when determining Qk_0 and especially E, due to the low degree of conversion (η_{ign}) attained for t_{ign}.

The determination of kinetic constants by the ignition method is considered in Sect. 5.4.

5.2 "Flash-Block" Device

A diagram of an installation used to study the kinetics of the high-temperature decomposition of blasting powders and composite solid propellants is shown in Fig. 5.1.

Aluminum block-heater *1* (cylinder 45 mm in diameter, 120 mm long) is equipped with resistance thermometer *2* that is attached to its lateral surface and a heating coil *3* (the temperature control system is similar to that of installation LP-3, so the other parts indicated in Fig. 5.2 are not listed). Sample *8* is placed onto a textolite plate *9* fixed onto lifter *10*, which is equipped with an anti-vibration buffer. In the experiments with nitrocellulose samples, thermostat *5* (a sealed steel vessel with inner channel *6*) was used to set various T_∞ values. Heat carrier *7* (water or carbon tetrachloride) is supplied to the thermostat through ultrathermostat connecting pipes. The temperature of the end of the bottom of the block is kept constant (±0.3°C).

The sample was lifted and pressed onto the block using a load fixed onto three threads that pass over pulleys and were connected to the lifter. The value of t_{ign} was measured (with good accuracy) using a stop-watch with a scale interval of 0.1 s. After sample inflammation, the kinematic unit was returned to the original position, and the block surface was cleaned from the residue (without cooling the block, using sandpaper wound onto a bush over cotton wool; the bush was connected to a high-speed motor via a flexible shaft, and the sandpaper did not burn onto the block due to the high rotation speed).

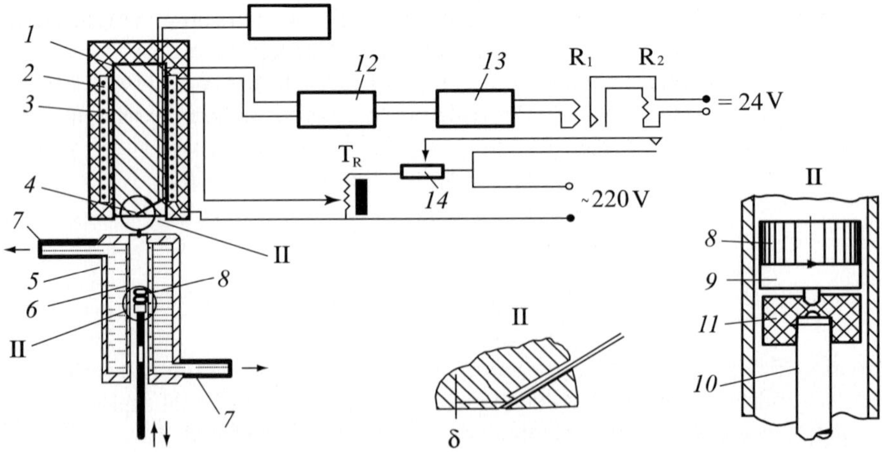

Fig. 5.1 Installation schematic: block heater (*1*), resistance thermometer (*2*), heating coil (*3*), hot thermojunction (*4*), thermostat body (*5*), thermostat channel (*6*), heat carrier (*7*), sample (*8*), textolite plate (*9*), lifter stem (*10*), lifter buffer (*11*), resistive bridge (*12*), amplifier (*13*), ballast rheostat (*14*)

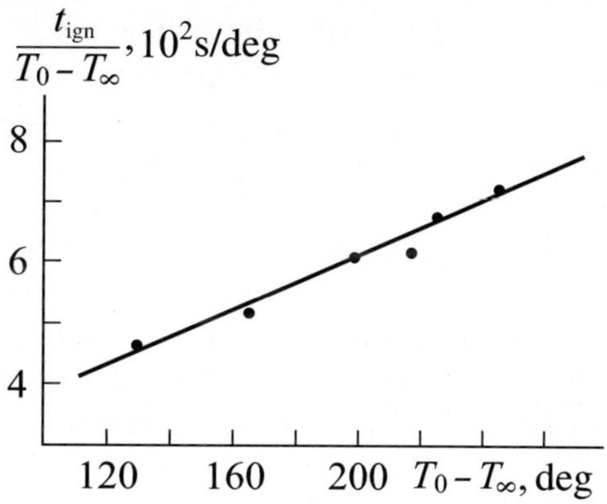

Fig. 5.2 Effect of the initial sample temperature, T_∞ on the ignition delay, t_{ign} ($T_0 =$ 498 K)

5.3 Studying the Kinetics of the High-Temperature Decomposition of Nitrocellulose

In the experiments [9], samples of nitrocellulose (which is well-characterized as a model system for condensed-phase reactions occurring without phase transitions [12]) were studied. The kinetics of the low-temperature decomposition of nitrocellulose has also been thoroughly investigated by classical isothermal methods (see for example [13]).

The samples were prepared in the form of pellets 12 mm and 18 mm in diameter and 6 mm long with a density of 1.49–1.50 g cm^{-3}. The block temperature was varied in the range of $T_0 = 485$–525 K, and the sample initial temperature in the range of $T_\infty = 255$–369 K. Before ignition the samples were kept in thermostat *6* (Fig. 5.1) for 30 min. This time was found to be enough for the complete preheating (cooling) of the sample up to the required temperature T_∞ (the increase in thermostating time did not influence t_{ign}). For the indicated ranges of T_0 and T_∞, t_{ign} was found to be in the range of 3–20 s.

Equation (5.5) is used to analyze the kinetic data. Noting that $\gamma \ll 1$ and $\beta \ll 1$, one can rewrite this expression as

$$\frac{Qk_0}{c} \exp\left(-\frac{E}{RT_0}\right) \frac{t_{ign}}{T_o - T_\infty} = 1.6 + 0.2\frac{E(T_0 - T_\infty)}{RT_0^2}. \tag{5.6}$$

It follows from Eq. (5.6) that for $T_0 =$ const and $T_\infty =$ variable, the dependence of $\frac{t_{ign}}{T_0 - T_\infty}$ on $(T_0 - T_\infty)$ should be linear. The experimental data obtained at $T_0 = 498$ K and $T_\infty = 255$–369 K are presented in the corresponding coordinates in Fig. 5.2.

The equation of the straight line obtained is

$$\frac{t_{ign}}{T_0 - T_\infty} = a + b(T_0 - T_\infty), \tag{5.7}$$

where $a = 1.6 \times 10^{-2}\,\mathrm{s\,deg}^{-1}$, $b = 2.3 \times 10^{-2}\,\mathrm{s\,deg}^{-2}$. However, the linearization of the experimental results in these coordinates does not completely prove the correctness of Eq. (5.6). The quantitative analysis of the parameters of the straight line appears to be of significant importance. The absolute values of these parameters that are strongly dependent on Qk_0, E and T_0 cannot be estimated from Eq. (5.6) with the required accuracy. In order to quantitatively compare the theoretical and experimental data, the ratio $\frac{a}{b} = 8\frac{\mathrm{R}T_0^2}{E}$, which is independent of Qk_0 and slightly (not exponentially) dependent on E and T_0, can be used. For $E = 202\,\mathrm{kJ\,mol}^{-1}$ [12], $\left(\frac{a}{b}\right)_{\mathrm{theor}} = 82°$ and $\left(\frac{a}{b}\right)_{\mathrm{exp}} = 70°$, which indicates a satisfactory agreement between the theoretical and experimental data.

The kinetic parameters Qk_0 and E can also be determined from the experimental data using Eq. (5.6). Let's rewrite this expression as

$$\lg\frac{t_{\mathrm{ign}}}{T_0 - T_\infty} = \lg\left[\frac{c}{Qk_0}\left(1.6 + 0.2\frac{E}{\mathrm{R}}\frac{T_0 - T_\infty}{T_0^2}\right)\right] + \frac{E\lg e}{\mathrm{R}}\frac{1}{T_0}. \tag{5.8}$$

If $\frac{T_0 - T_\infty}{T_0^2}$ is kept constant in the experiments, a plot of $\lg\frac{t_{\mathrm{ign}}}{T_0 - T_\infty}$ against T_0^{-1} is linear, and E can be found from the slope ratio and Qk_0 from the intercept on the y-axis. Under regular experimental conditions, $\frac{T_0 - T_\infty}{T_0^2}$ changes only slightly; for example, in the case of experiments with nitrocellulose samples at $T_\infty = 293$–$298\,\mathrm{K}$ and $T_0 = 485$–$515\,\mathrm{K}$, it varied by 5%. Thus, this approach can be used to analyze experimental data obtained when $T_0 \approx$ const and $T_\infty =$ variable. This is very convenient, since if the initial sample temperature is equal to the room temperature, an additional thermostat (*5*, Fig. 5.1) is not needed. Experimental data obtained for the indicated range of T_0 are presented in Fig. 5.3 (each point corresponds to a t_{ign} value that is the average from three/four experimental runs). The good agreement between the data obtained for samples of different diameters implies that t_{ign} is independent of the lateral heat losses. Analysis of the data shown in Fig. 5.3 using Eq. (5.8) gave

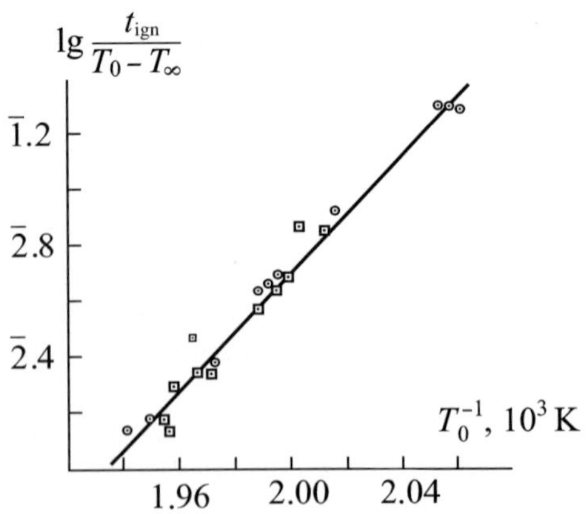

Fig. 5.3 Arrhenius anamorphosis presentation of the experimental data showing the dependence of the ignition delay, t_{ign}, on the block temperature, T_0, at $T_\infty = 293$–$298\,\mathrm{K}$ for the samples 12 mm (*circles*) and 18 mm (*squares*) in diameter

$E = 200\,\mathrm{kJ\,mol^{-1}}$ and $Qk_0 = 10^{23}\,\mathrm{J\,g^{-1}\,s^{-1}}$. The values obtained are in a good agreement with the literature data on k_{02} and E_2 for the decomposition of nitrocellulose [12, 13, 14, 15]. The results of kinetic studies of nitrocellulose are summarized in Table 8.1, Chap. 8.

The kinetic constants of the limiting reaction of high-temperature decomposition of ammonium bichromate at $T_0 = 518$–$559\,\mathrm{K}$ ($E = 166\,\mathrm{kJ\,mol^{-1}}$, $Qk_0 = 8.3 \times 10^{17}\,\mathrm{J\,g^{-1}\,s^{-1}}$) and organic azide–APU at $T_0 = 525$–$619\,\mathrm{K}$ ($E = 180\,\mathrm{kJ\,mol^{-1}}$, $Qk_0 = 10.4 \times 10^{18}\,\mathrm{J\,g^{-1}\,s^{-1}}$) were determined from ignition experiments using the "flash-block" device. The data obtained for organic azide–APU are in good agreement with the corresponding data for low-temperature ($T_0 = 420$–$489\,\mathrm{K}$) decomposition: $E = 175\,\mathrm{kJ\,mol^{-1}}$, $Qk_0 = 6.2 \times 10^{18}\,\mathrm{J\,g^{-1}\,s^{-1}}$ (up to $\eta \approx 0.8$). This confirms the applicability of the flash-block technique to the study of the initial stage kinetics of fast exothermal reactions in homogeneous condensed systems.

5.4 Studying the Initial Stage Kinetics for the High-Temperature Decomposition of Model and Commercial Composite Solid Propellants

5.4.1 Method Verification for the Ignition of Polyurethane-Based Composite Solid Propellant

The decomposition of polyurethane-based (grade P) composite solid propellant was the first process studied using the flash-block device due to availability of some data on the low-temperature decomposition kinetics of this system [16] and a few experimental data on the kinetics of heat release (obtained thermographically in a hot gas flow) [17]. Since it is a typical representative of commercial AP-containing composite solid propellants (based on its composition and ignition patterns), the polyurethane-based energetic material was found to exhibit ignition behavior qualitatively similar to that of a number of commercial and model "AP–fuel binder" systems.

The samples (10 mm in diameter, 6 mm long) were chopped from hollow rods made by a standard molding method using a special cutter (ensuring strictly parallel and smooth end-surfaces). The sample composition is shown in Table 5.1.

Table 5.1 Composition of the polyurethane-based (P) fuel

Component	wt %
Monodisperse AP	70.86
Polyester	18.84
Toluylenediisocyanate	2.8
Aluminum powder	7.0
Lithium fluoride	0.5

As T_0 increases from 583 K to 665 K, t_{ign} decreases from 77 s to 2.2 s. In the plot of $\lg \frac{t_{ign}}{T_0 - T_\infty}$ against $\frac{1}{T_0}$, two qualitatively different regions were found to exist. For T_0 in the range of 583–628 K, the experimental data are well linearized down to $t_{ign} \approx 4$ s, and a strong dependence of t_{ign} on T_0 is apparent (Fig. 5.4a). When T_0 is increased further, the change in t_{ign} becomes less pronounced, attaining an approximately constant value of 2.2–2.3 s. It is interesting that in the second region isolated clicks, sometimes accompanied by sparkling in the block–sample contact, were heard before ignition. These phenomena did not result in sample ignition. However, the plotted points corresponding to the detection times of these phenomena (Fig. 5.4a) were found to lie on the extension to the experimental straight line (though with noticeable scatter). These allow one to draw a conclusion about the inapplicability of the homogeneous thermal ignition model in the second region due to reaction layer thinning and the dominance of heterogeneous effects, whereas in the first region this model is quite valid. Estimating the critical thickness of the reaction layer for $t_{ign} \approx 2.2$ s using Eq. (5.4) gives $\Delta_{ign} \approx 8.3 \times 10^{-5}$ m, which is very close to the size of a single AP crystal in the polyurethane-based fuel (10^{-4} m).

After determining the applicability limits for the homogeneous thermal ignition model, one finds from the slope of the dependence (Fig. 5.4a) that $E = 200 \pm 8\,\mathrm{kJ\,mol^{-1}}$. This value can then be substituted into Eq. (5.8) for the further analysis of the data using $\lg[Qk(T_0)] - 1/T_0$ coordinates (the second approximation, see Fig. 5.4b). The heat capacities of most composite solid propellants are in the range 2–2.25 $\mathrm{J\,g^{-1}\,deg^{-1}}$; for the polyurethane-based (P) fuel $c = 2.1\,\mathrm{J\,g^{-1}\,deg^{-1}}$.

For $T_0 = 582$–626 K, the values of E and Qk_0 were found to be $200 \pm 8\,\mathrm{kJ\,mol^{-1}}$ and $4 \times 10^{19.0 \pm 0.7}\,\mathrm{J\,g^{-1}\,s^{-1}}$, which are close to the results obtained by thermography of the decomposition of the polyurethane-based (P) material in a hot gas flow [17].

It was interesting to compare the results obtained with literature data on the kinetics of low-temperature decomposition that can be described [16] by the autocatalytic-type equation

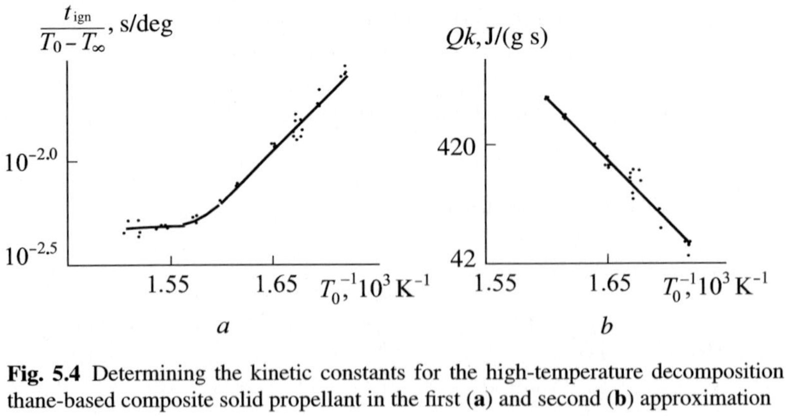

Fig. 5.4 Determining the kinetic constants for the high-temperature decomposition of polyurethane-based composite solid propellant in the first (**a**) and second (**b**) approximation

$$\frac{d\eta}{dt} = k_1(1-\eta) + k_2\eta^{0.5}(1-\eta), \tag{5.9}$$

where

$$k_2 = 10^{8.5}\exp\left(-\frac{28000}{RT}\right)\left[s^{-1}\right] \tag{5.10}$$

for temperatures lower than or equal to 493 K and

$$k_2 = 10^{12.8}\exp\left(-\frac{38000}{RT}\right)\left[s^{-1}\right] \tag{5.11}$$

for temperatures higher than 493 K.

The kinetic constants for low-temperature decomposition in Eqs. (5.10) and (5.11) significantly differ from that for high-temperature decomposition, as determined from experimental ignition data (here $Q = 2.08\,\mathrm{kJ\,g^{-1}}$, from [17], was used):

$$k_{\mathrm{high-temp}} \approx 2\cdot 10^{16}\exp\left(-\frac{48000}{RT}\right)\left[s^{-1}\right]. \tag{5.12}$$

An analytical expression for k_1 could not be derived in [18] due to the extremely strong autocatalytic effect ($k_1/k_2 \ll 1$). The approximate order of magnitude for k_1 was found from the results of specially designed experiments (that measured the weight decrease under isothermal conditions) to be $10^{-5}\,\mathrm{s}^{-1}$ at $T = 493$ K. Extrapolation to $T = 493$ K using Eq. (5.12) gives a similar constant value of $1.5\times 10^{-5}\,\mathrm{s}^{-1}$. Taking into account the long range of the extrapolation, one can consider this agreement between the data to be quite satisfactory.

Thus two preliminary conclusions can be drawn:

(a) The kinetic constants of the limiting stage of the fast high-temperature decomposition of composite solid propellants can be evaluated (for a certain range of t_{ign}) by the ignition method using the flash-block device.

(b) The satisfactory agreement between the k_1 data for high-temperature and low-temperature decomposition of the polyurethane-based (P) material obtained for high E (determined by the ignition method) may not be just a random coincidence. In this case, reactions with such high values of E are assumed to occur at low temperatures as well. Due to high activation barriers, the contributions from these reactions are not noticeable at significant degrees of conversion in the low-temperature region, while at higher temperaturesthey become more important. Correspondingly, the effective constants of high-temperature decomposition significantly differ from those in the low-temperature region.

Thus the ignition method can be used to evaluate the kinetic constants for reactions that dominate in the high-temperature region (if the decomposition mechanism is constant over a wide temperature range). The low-temperature kinetics of these reactions cannot be studied by traditional techniques due to their negligibly small contributions to the overall process.

5.4.2 High-Temperature Decomposition of Model and Commercial Composite Solid Propellants

The kinetics of the high-temperature decomposition of a number of model and commercial composite solid propellants (including standard and fast-burning materials based on rubber resin and polyurethane fuel-binders and AP) were studied by the ignition method.

In the experiments, fuel–AP model materials as well as multicomponent systems (containing aluminum and combustion catalysts: Fe_2O_3 and $CuCr_2O_7$) were investigated. The compositions of the studied systems and the kinetic data obtained are presented in Table 5.2 (the plots and calculations similar to those given in Sect. 5.4.1 are omitted).

The ignition behavior exhibited by these systems was found to be similar to that of the polyurethane-based material. For the systems with a high content of the fine AP fraction, there is no inflection on the $\lg \frac{t_{ign}}{T_0 - T_\infty}\left(\frac{1}{T_0}\right)$ curves down to $t_{ign} \approx 1$ s, which implies that the thermal decomposition has a quasi-homogeneous character. For systems with a lower content of the fine AP fraction (and in the polyurethane-based fuel), a change in the ignition mechanism was found to occur at $t_{ign} \approx$ 2.0–2.5 s.

5.4.3 Final Remarks

Comparative analysis of the kinetic data obtained for systems with various binders, catalysts and additives resulted in the following conclusions:

1. The effective kinetic constants (E and Qk_0) of the catalyst-free high-temperature decomposition of condensed materials differ significantly from those associated with low-temperature decomposition (determined by isothermal and thermal explosion methods).
   For example, the activation energies for the polyurethane- and chloropolyurethane-based fuels (NG-3-63H and NG-2-60-DTM), determined from the critical conditions of the thermal explosion at $T = 413$–473 K, are in the range 117–133 kJ mol^{-1}, while at higher temperatures (see Table 5.2) they are approximately 170 kJ mol^{-1} for the polyurethane-based fuel and 210 kJ mol^{-1} for the chloropolyurethane-based one. Materials with polyolefin and some other binders (bitumen, polyethylene, rubber resin) are also characterized by high activation energies for high-temperature decomposition. This suggests that the macrokinetics of the high-temperature decomposition of AP-based materials are determined by kinetic constants that are different from those for the low-temperature decomposition (this aspect is also discussed in Chap. 9).
2. The addition of aluminum hardly influences the high-temperature decomposition of AP-based materials.

Table 5.2 Kinetic parameters for the initial stages of the high-temperature decomposition of AP-based composite solid propellants obtained by using the flash-block device

Composite solid propellant composition			Temperature, T_0, K	Ignition delay, t_{ign}, s	Activation energy, E, kJ mol^{-1}	$\lg(Qk_0)$, J g^{-1} s^{-1}
Composition grade (binder)	Metal	Catalyst				
NG-2-60-DTM	none	none	641–735	3.4–68	178 ± 6	16.4 ± 0.4
(Polyurethane–DTM)	none	Fe_2O_3	576–692	1.2–87	129 ± 2	13.2 ± 0.2
	none	$CuCr_2O_7$	634–693	8.0–102	162 ± 3	14.8 ± 0.3
	Al	none	661–765	0.8–37	160 ± 3	14.8 ± 0.2
	Al	Fe_2O_3	590–673	2.6–55	121 ± 2	12.3 ± 0.2
	Al	$CuCr_2O_7$	636–707	3.4–94	172 ± 2	15.8 ± 0.2
	Al	SiO_2	651–706	2.0–30	163 ± 12	14.6 ± 1
NG-3-63H	none	none	624–671	4.6–100	204 ± 8	19 ± 0.8
(Chloropolyurethane)	none	Fe_2O_3	554–655	1.0–154	157 ± 2	16.1 ± 0.1
	Al	none	624–677	2.4–65	213 ± 6	19.6 ± 0.5
	Al	Fe_2O_3	567–664	1.0–84	143 ± 3	15.9 ± 0.5
NG-3-66A	Al	none	592–643	6.1–70	143 ± 2	14.3 ± 0.5
(Organic azide–APU)	Al	Fe_2O_3	548–616	6.0–125	127 ± 3	13.5 ± 0.2
	Al	$CuCr_2O_7$	563–627	9.0–100	121 ± 4	12.5 ± 0.4
SKT (Rubber resin–S)	Al	Fe_2O_3	696–774	1.3–91	242 ± 6	20 ± 0.4
S-2 (Rubber resin–P)	Al	Fe_2O_3	639–737	0.8–83	186 ± 4	16.9 ± 0.3
80 wt% AP + 20 wt% polyethylene	Al	Fe_2O_3	667–707	2.8–33	233 ± 12	20.6 ± 1
80 wt% AP + 20 wt% bitumen	Al	Fe_2O_3	619–701	1.3–68	242 ± 8	21.4 ± 0.4
P (Polyurethane–P)	Al	Fe_2O_3	582–626	4.3–77	200 ± 8	19.6 ± 0.74

3. Combustion catalysts such as $CuCr_2O_7$ and SiO_2 were not found to affect the high-temperature decomposition of the material's condensed phase. The addition of Fe_2O_3 results in a significant acceleration of the high-temperature reaction and a decrease in the activation energy value down to $E \approx 125\,\mathrm{kJ\,mol^{-1}}$.
   Joint analysis of these results and data on low-temperature decomposition and thermal explosions, as well as results from experiments on the effect of Fe_2O_3 and $CuCr_2O_7$ on materials burning in the chemical arc, have led to the following conclusion. In contrast to Fe_2O_3, which catalytically affects high-temperature processes in the condensed and gas phases of the burning fuel, adding $CuCr_2O_7$ to the binder mainly influences gas-phase reactions (the introduction of $CuCr_2O_7$ into AP pellets is known to result in accelerated decomposition and burning). However, in composite solid propellants, as a rule there is no a direct contact between the AP grains and the catalyst (the latter being uniformly distributed in the binder volume which occupies intergranular space in the composite). As opposed to other ignition techniques (radiant flux, electrically heated wires, etc.), ignition by means of a hot heat-conducting block provides information about the kinetics of reactions in the condensed phase. This is explained by the almost complete suppression of the release of heat from the gas-phase reaction in the thin gap between the sample and the block due to the "cooling" action of the latter.
4. The kinetic constants for slow-burning polyolefin-based model systems and for fast-burning rubber resin-based material are quite similar. In this case, one can assume that the high burning rate of the latter system is associated with the high degree of homogeneity of the material due to the use of submicron grain-sized AP. As a result, the degree of conversion of the condensed phase attains quite a high value due to the high-temperature thermally self-accelerating reaction ($E = 242\,\mathrm{kJ\,mol^{-1}}$) before the rate of volumetric heat release in the fuel surface layer becomes limited by dispersion and microheterogeneity effects.
5. The simple and reliable flash-block-based ignition technique can be used as an express method for evaluating high-temperature decomposition parameters and determining the ignition delay times for composite solid propellants under fixed thermophysical conditions (intensive aerodynamic heating and others).

References

1. Zel'dovich JaB (1939) Zh Eksp Teor Fiz 9:1530
2. Merzhanov AG (1965) In: Encyclopaedia on Physics. Sovetskaya entsiklopediya, Moscow, p 458
3. Merzhanov AG, Dubovitsky FI (1966) Usp Khim 35:656
4. Merzhanov AG, Averson AE (1971) Combust Flame 6:89
5. Averson AE (1968) PhD thesis. Institute of Chemical Physics of the USSR Academy of Sciences, Chernogolovka, Russia
6. Rosenband VI, Barzykin VV (1967) Combust Explos Shock Waves 4:283
7. Rosenband VI, Barzykin VV (1967) Combust Explos Shock Waves 4:96
8. Carslaw HS, Jager JC (1959) Conduction of heat in solids, 2nd edn. Clarendon, Oxford
9. Shteinberg AS, Ulybin VB, Barzykin VV, Merzhanov AG (1966) J Eng Phys Thermophys 4:282

10. Mikheev MA (1956) Fundamentals of heat transfer. GEI, Moscow
11. Averson AE, Barzykin VV, Merzhanov AG (1965) J Eng Phys Thermophys 2:172
12. Grigor'ev YuM, Maksimov EI, Merzhanov AG (1965) Combust Explos Shock Waves 1:70
13. Manelis GB, Rubtsov YuI, Smirnov LP, Dubovitsky FI (1962) Kinet Catal 1:32
14. Lisitsky VI (1970) Combust Explos Shock Waves 6:494
15. Grigor'ev YuM, Lisitsky VI (1967) Combust Explos Shock Waves 3:313
16. Manelis GB, Nazin GM, Rubtsov YI, Strunin VA (2003) Thermal decomposition and combustion of explosives and propellants. Taylor & Francis, New York
17. Lisitsky VI, Merzhanov AG (1965) Combust Explos Shock Waves 1:45
18. Manelis GB, Rubtsov YuI (1968) Kinetics of thermal decomposition of composite solid propellant "P". Report of the Institute of Chemical Physics of the USSR Academy of Sciences, Chernogolovka, Russia

Conclusion for Part I

The potential value of the experimental and theoretical results discussed in Part I for the further development of ideas about the nature of the high-temperature decomposition of energetic materials (in particular, composite solid and hybrid propellants) is a matter of significant interest.

The spatially nonisothermal techniques of ignition and linear pyrolysis (which allow one to obtain a certain essentially constant characteristic of the studied system, t_{ign} or U, respectively) are considered to be "discrete" methods according to a classification suggested in [1]. In the experiments the rates of the overall processes are believed to be quite close to those of processes that occur in real solid-propellant rocket engines and hybrid rocket engines. For example, the rate of polymer linear pyrolysis performed in the LP installations reached a value of 10^{-1} cm s^{-1}. This is quite close to the burning rates of hybrid propellants and only an order of magnitude lower than the linear pyrolysis rates characteristic of composite solid propellant binders. The existence of the "third mode" of polymer linear pyrolysis (found for PMMA at $U > 5 \times 10^{-2}$ cm s^{-1}) allows one to propose that similar macrokinetic patterns occur for other polymeric fuel binders: a weak decrease in T_S at significantly increased linear pyrolysis rates. The burning of polymer spheres was found to obey the Sreznevsky law (the linear reduction in the surface area of a burning polymer particle as a function of time: $d^2 = d_0^2 - \kappa t$), which indirectly confirms the existence of the third mode for fast linear pyrolysis. This leads to the resolution of the contradiction between data on high-temperature decomposition kinetics obtained in early works on polymer linear pyrolysis [2, 3, 4] and burning rates and surface temperatures as well as temperature profiles found for the condensed phases of polymer–AP systems [5, 6, 7, 8, 9, 10]. Extrapolation of these experimental data (by presenting the dependence of the linear pyrolysis rate on the surface temperature in the form of an Arrhenius plot) to $U \approx 10^{-1} - 1$ cm s^{-1} resulted in unrealistically high T_S values. In contrast, the T_S values determined for the third mode agreed with the data obtained by microthermocouple measurements for the burning of bulk PMMA in flowing oxygen [6, 11] and AP–PMMA mixtures [7].

In the chemical arc, linear pyrolysis of the components takes place in the flame of interaction between the gaseous products of the oxidizer and fuel-binder decomposition. Under these conditions, the burning occurs in one of its limiting modes (in its "elementary act," to be exact), which is an important stage in the complex process of heterogeneous system burning. Under the conditions in the chemical arc, there is no interaction between the condensed phases of the components. Therefore they can only mutually influence the linear pyrolysis rates via the gas-phase processes. The combustion of AP and PMMA linear pyrolysis products was found to be nonstoichiometric (a significant excess of the fuel). The temperature in the gas gap attains its maximum value near the fuel surface. The role of catalysts at various burning stages can be clarified experimentally by introducing catalysts into an oxidizer or a fuel. The combination of the chemical arc method with other techniques of studying limiting burning modes (for example, the method based on the burning of "sandwich" systems, etc. [6, 12]) seems to be a very promising approach for further investigations of the mechanism of composite solid propellant burning.

Thus linear pyrolysis methods can be used to study the macrokinetics of high-temperature reactions in condensed phases of solid fuel components under purely thermal influence and under the effect of the interaction flame. Pyrolysis of the fuel is a very important stage in the burning of composite solid propellants, and especially hybrid propellants. Nowadays, ensuring high linear rates of fuel burning is one of the main goals in the development of promising environment-friendly hybrid propellants and hybrid rocket engines. In this regard, the experiments on the linear pyrolysis of linear and crosslinked PMMA seem to be quite informative. The high-temperature decomposition occurring in the third mode was found to be affected by the rate of heat supply (q_S) to the pyrolysis surface. In this case, the total gasification heat (Q_Σ) is the main characteristic determining the rate of linear pyrolysis (at a certain q_S). Q_Σ includes the thermal effect of pyrolysis and the heat of the material warm-up from T_∞ to T_S. This parameter can be considered a high-temperature decomposition constant characteristic of the particular material since the surface temperature changes only slightly in the third mode.

Values of Q_Σ were measured by the block-calorimeter method for a number of model and commercial fuels. The estimation of Q_Σ from thermodynamic data is very difficult, since, as a rule, information on linear pyrolysis product composition, the value of T_S and the temperature dependence of the heat capacity of the fuel is not available. At the temperatures and times characteristic of linear pyrolysis the processes in the reaction zone cannot attain thermodynamic equilibrium. The Q_Σ values were analyzed and compared with data on the rates of linear burning in a model hybrid rocket engine chamber. The results of this analysis suggest a general behavior pattern even for fuels that are significantly different in terms of their structures and low-temperature decomposition parameters. Thus the following conclusions regarding the mechanism of pyrolysis and burning of fuels in a hybrid rocket engine can be made:

- Linear pyrolysis of fuel occurs in the third mode (the influence of purely physical factors associated with hydrodynamics in a hybrid rocket engine on the rate of hybrid propellant burning is not considered here). The polymer fuel burning pattern is weakly affected by the kinetics of reactions in the condensed phase and on the pyrolysis surface.
- Decreasing the fuel's Q_Σ leads to a significant increase in the rate of hybrid propellant burning. This can be done by, for example, adding homogeneous

compounds that decompose with high heat release in the condensed phase (organic azides).
- The rate of fuel burning in a hybrid rocket engine is affected by the pressure, because the latter influences the macrokinetics of gas-phase processes.
- The burning of the hybrid propellant is limited mainly by heat transfer in the interface layer rather than by the kinetics of linear pyrolysis in the condensed phase. This explains the effect of Q_Σ on the fuel burning rate.

Comparison of experimental data on Q_Σ with the burning rate data obtained under identical conditions for mixtures of polymers or individual organic compounds with AP did not suggest any correlations between them [13]. This result is quite important for understanding the mechanism of composite solid propellant burning. The reactivities of the components and the products of their high-temperature decomposition are more important factors for the burning of composite solid propellants than the thermal effect of binder linear pyrolysis. In other words, rates of heat- and mass-transfer when burning composite solid propellants (heterogeneous mixtures with characteristic structural dimensions of fractions of mm) are close to or exceed the rates of corresponding chemical reactions. The role of the gas-phase and condensed-phase reactions in the overall process can be determined by direct kinetic studies of the burning materials. As far as we know, the kinetics of gas-phase reactions between products of the high-temperature linear pyrolysis of AP and practically interesting polymers has not been investigated experimentally. However, according to the results of a number of experimental and theoretical studies on the burning of AP-based composite solid propellants, condensed-phase reactions can also play a significant role [14, 15, 16].

Despite some limitations (see below), the application of the flash-block ignition method allows one to determine the constants of the high-temperature decomposition of numerous model and commercial composite solid propellants at the preheat and reaction layer temperatures for the first time. Classical isothermal methods cannot be used to determine these constants in the high-temperature region (573–723 K). This is mostly due to the huge difference between the effective kinetic parameters (E, k_0) that determine the rates of composite solid propellant decomposition at low and high temperatures. It was found for polyurethane-based material (and some others; for example, rubber resin-based composite solid propellant) that the contribution from reactions that dominate at high temperatures to low-temperature decomposition is negligibly small, even when the formal kinetics of system decomposition does not change over a wide temperature range. In this case, any low-temperature data-based extrapolations into the high-temperature region are almost impossible.

The studies considered in this book were followed by numerous works that confirmed the correctness of the methods and the conclusions obtained using them [1, 10, 17, 18, 19, 20, 21, 22, 23, 24, 25, 26, 27, 28, 29]. For example, Shlensky experimentally determined that the forced high-temperature decomposition of a number of polymers occurs in the third mode; that is, at an approximately constant temperature for a wide range of decomposition rates [30]. The main conclusions drawn regarding the ignition of homogeneous and heterogeneous solid energetic materials were confirmed by Muratov and coworkers [31].

To obtain quantitative data on the kinetics of fast condensed-phase reactions, special experimental and analytical approaches should be used due to the spatial and (or) temporal nonisothermal nature of the material. The linear pyrolysis and ignition methods are based on such approaches and allow one to obtain the required data using quite simple experimental setups. However, these techniques have serious limitations.

Since they are "discrete" techniques [32], the linear pyrolysis and ignition methods cannot be used to derive an analytical expression that describes the formal reaction kinetics directly from experimental data. In order to estimate E and Qk_0 or E and k_0 using the ignition and linear pyrolysis methods, the corresponding expression for $\varphi(\eta)$ in the equation $\dot{\eta} = k(T)\varphi(\eta)$ (describing the dependence of the reaction rate of the conversion degree, η) should be assigned a priori. If the formal kinetics of the self-accelerating (under isothermal conditions) reaction can be described by, for example, expression $\varphi(\eta) = (1-\eta)^n$, an error in the determination of the constants associated with the uncertainty of the reaction order n is not that significant (in the experiments on linear pyrolysis, decomposition can be considered a "pseudo-zero-order" process due to the densification of the reaction layer by the force pressing the sample onto the heater; during regular (nondegenerative) ignition in the flash-block device the degree of conversion in the surface layer by time t_{ign} is small [33]). However, due to this limitation, neither method can be considered comprehensive. For many homogeneous and heterogeneous systems, the degree of decomposition attained during burning is quite high. Therefore, the availability of an analytical expression for the dependence $\dot{\eta} = \dot{\eta}(\eta, T)$ at high temperatures is extremely important for identifying the burning mechanism.

Another disadvantage of the linear pyrolysis and ignition methods (although less serious) is their pretty limited applicability. The method of linear pyrolysis on a solid heater can only be used for systems that decompose without forming a significant amount of condensed residue (ammonium and nitronium compounds, some polymers and organic compounds).

The flash-block method can be successfully applied to the study of systems that decompose at high temperatures with significant heat release in the condensed phase and without (at least at low conversion degrees) vigorous gas evolution. As a rule, decomposition of these systems results in the formation of a condensed residue.

Both methods can be used to study systems that burn without phase transformations. However, the decomposition of numerous practically interesting materials, particularly many components of solid and liquid propellants, are accompanied by phase transformations (sublimation of solid oxidizers, boiling and evaporation of liquid monopropellants), even at moderate temperatures. Thus, in order to correctly identify the ignition and burning mechanisms, the quantitative evaluation and separate analysis of the contributions of physical and chemical processes to the overall process macrokinetics are of great importance.

For example, when the experiments described here were started, a fundamental contradiction between the conclusions drawn by Belyaev and Merzhanov regarding the mechanism of the thermal self-ignition of volatile explosives had not been resolved due to a lack of information about the role of the phase transformation heat in the heat balance. According to data obtained by Belyaev [5], the ignition of volatile explosives takes place in the gas phase. Therefore, the critical conditions for this process are determined by the kinetics of the decomposition of the explosive vapors. In contrast, Merzhanov and coworkers showed [34] that for typical secondary explosives such as Tetryl and DINA the critical conditions and the ignition delay period are determined by constants of the condensed-phase reactions.

The study of the macrokinetic features of the decomposition of such systems is of significant practical and theoretical interest, especially at high temperatures, when the saturated vapor pressure and the ambient pressure become commensurable. Promising monopropellants for liquid rocket engines and torpedo power units, such as highly concentrated hydrogen peroxide, anhydrous hydrazine, propylene glycol dinitrate-based monopropellant and some other commercial energetic materials, can also be considered volatile explosives due to their decomposition and combustion patterns. The macrokinetic patterns of the high-temperature decomposition of AP and volatile explosives have some similar features (AP is prone to exhibiting equilibrium sublimation in parallel with irreversible thermal decomposition). However, this similarity is more pronounced for liquids such as highly concentrated hydrogen peroxide, organic nitrates and hydrazine.

The potential for the application of monopropellants as explosives that ignite upon contact with hot walls in military airplane quick-firers has been actively discussed in publications over the past few years. It is obvious that before designing and building machines of this kind, as well as developing safety protocols, the decomposition macrokinetics of the materials used should be thoroughly studied over the required temperature and pressure range.

Part II of this monograph is devoted to the study of the high-temperature decomposition kinetics of solid and liquid propellants under the conditions of spatially isothermal heating. The macrokinetic aspects of thermal decomposition and explosion of volatile explosives and monopropellants are also considered.

References

1. Shchukin VK, Reshetnikov SM (1973) Trans Kazan' Inst Aviat 154
2. Schultz RD, Dekker AO (1955) In: Proceedings of the 5th Symposium on Combustion. Reinhold, New York
3. Chaiken RF (1962) J Chem Phys 37:2311
4. Chaiken RF, Andersen WH, Barsh MK (1960) J Chem Phys 32:141
5. Bakhman NN, Belyaev AF (1967) Combustion of heterogeneous condensed systems. Nauka, Moscow
6. Chorpening BT, Knott GM, Brewster MQ (2000) Proc Combust Symp 28:847
7. Zenin AA (1976) DSci thesis. Institute of Chemical Physics of the USSR Academy of Sciences, Moscow
8. Powling J, Smith WA (1962) Combust Flame 6:173
9. Powling J, Smith WA (1963) Combust Flame 7:269

10. Ishihara A, Sarai Y, Konishi K, Andoh E (2005) Proc Combust Inst 30:2123
11. Shteinberg AS, Ulybin VB, Manelis GB, Dolgov EI (1972) In: Stesik LN (ed) Combustion and explosion. Nauka, Moscow, p 124
12. Bachman NN, Librovich VB (1970) Combust Flame 15:143
13. Braginsky VI, Bakhman NN (1971) Combust Explos Shock Waves 7:240
14. Manelis GB, Strunin VA (1972) In: Stesik LN (ed) Combustion and explosion. Nauka, Moscow, p 53
15. Manelis GB, Strunin VA (1971) Combust Flame 17:69
16. Manelis GB, Nazin GM, Rubtsov YI, Strunin VA (2003) Thermal decomposition and combustion of explosives and propellants. Taylor & Francis, New York
17. Alekseev YuI, Korolyov VL, Knyazhitsky VP (1972) In: Stesik LN (ed) Combustion and explosion. Nauka, Moscow, p 128
18. Rusanov VP Popov NP (1971) Teplofiz Vys Temp 1:9
19. Rusanov VP Popov NP (1971) Teplofiz Vys Temp 2:40
20. Reshetnikov SM (1975) PhD thesis. Kazan' Institute of Aviation, Kazan'
21. Reshetnikov SM (1975) Trans Kazan' Inst Aviat 184
22. Kirsanov YuA, Shchukin VK, Reshetnikov SM (1975) Combust Explos Shock Waves 10:15
23. Shchukin VK, Dresvyannikov FN, Kirsanov YuA (1973) College Trans Aviat Eng 4
24. Kirsanov YuA (1973) Trans Kazan' Inst Aviat 158
25. Kirsanov YuA (1973) PhD thesis. Kazan' Institute of Aviation, Kazan'
26. Gainutdinov RSh, Enaleev RSh (1972) In: Stesik NL (ed) Combustion and explosion. Nauka, Moscow
27. Alekseev YuI (1972) PhD thesis. State Institute of Applied Chemistry, St. Petersburg, Russia
28. Khalturinsky NA, Popova TV, Berlin AA (1984) Rus Chem Rev 53:197
29. Khalturinsky NA, Berlin AA (1990) Int J Polym Mater 14:109
30. Shlensky OF, Aksenov LN, Shashkov AG (1991) Thermal decomposition of materials: effect of highly intensive heating. Elsevier, Amsterdam
31. Muratov SM, Makharinsky VM, Afanas'ev GT (1972) In: Stesik LN (ed) Combustion and explosion. Nauka, Moscow, p 30
32. Merzhanov AG (1973) Combust Explos Shock Waves 9:4
33. Averson AE, Barzykin VV, Merzhanov AG (1965) J Eng Phys Thermophys 2:172
34. Merzhanov AG, Dubovitsky FI (1966) Usp Khim 35:656

Part II
Study of the Macrokinetics of High-Temperature Decomposition of Condensed Materials Under Spatially Isothermal Heating

Chapter 6
Mechanical and Thermal Dilution Methods Used in the Thermographic and Thermogravimetric Analysis of High-Temperature Decomposition of Condensed Materials

Abstract The need to substantially modify common nonisothermal methods such as TGA, DTA and DSC when used to study the kinetics of fast exothermal and endothermic reactions in condensed media is theoretically demonstrated. Modified methods based on deep ballasting of the sample with inert heat-conducting material are proposed. For homogeneous samples, "mechanical dilution" (mixing the sample particles with chemically inert material in the ratio 1:100) is used. For heterogeneous samples, a "thermal dilution" method where a thin sample is pressed between the ends of two metal cylinders characterized by high thermal conductivity is used. An analytical approach for determining kinetic constants from TGA, DTA and DSC data is proposed. Use of the modified experimental cells was shown to increase the upper limit of the reaction rates that could be measured by several orders of magnitude, even when commercial TGA, TDA and DSC instruments are applied.

6.1 Introduction

Although they have been widely used in physicochemical studies for a long time, differential thermal analysis (DTA, thermography), differential scanning calorimetry (DSC) and thermogravimetric analysis (TGA, thermogravimetry) have only recently been applied to obtain kinetic data (the most comprehensive information on the theory and practice of DTA can be found in [1, 2, 3, 4, 5, 6]).

Compared to isothermal methods, DTA and TGA are more informative with respect to the kinetic behavior of the system under study. In addition, the corresponding equipment is quite simple. The application of DTA and TGA to study fast high-temperature reactions in condensed media appears to be very promising.

When the experiments described here were begun, almost all methods of determining kinetic parameters from DTA–TGA data (except for a method used in [7]) were "discrete" ones [1]. As a rule, only some points on the experimental curve were used for analysis: peak height, maximum reaction rate, furnace temperature corresponding to these points, etc. When most or all of the experimental points were used for the analysis, the analytical expression describing the kinetics of the system was assigned a priori, which is characteristic of a "discrete" technique. For these methods, an analytical approach is generally considered to be correct if the experimental data can be linearized satisfactorily (additional information regarding the fundamental limitations of this approach can be found in [5]). An evident drawback of this technique is the "discrete" nature of the analytical approach, since only a small part of the "continuous" information contained in the set of DTA and TGA curves is used.

The use of experimental methods based on the continuous recording of the measured characteristics (including all methods based on a programmed heating schedule) allows one to obtain more kinetic information and to find an analytical expression that describes the studied process kinetics without any a priori assignments. Historically DTA and TGA equipment were designed for qualitative chemical analysis. Therefore, commercially available devices were not always very convenient for experimental studies of kinetics, particularly in relation to problems associated with setting up controlled thermophysical conditions in the reaction cell, which are required to obtain a correct analytical description of the process. Quite frequently this aspect was neglected under the assumption that one can obtain the required quantitative experimental data by decreasing the heating rate and the sample weight. However, in many cases, these measures were found to be insufficient [8].

As a rule, the analytical approach used to determine the kinetic parameters of a nonisothermal process from experimental data is based on the solution of the direct kinetic problem related to the development of the process. In the case of thermography (and similar techniques), this involves the programmed heating of the reacting material. This problem can be solved by employing various assumptions that determine the correctness of the applied analytical approach. The analytical approaches used can be divided into three groups.

a) In most works, the linear heating problem (the linear time dependence of the temperature) is solved by using only one kinetic equation (see for example [9, 10]). This approach is quite popular due to the simplicity of the computations involved. In this case, the main drawback is the complete neglect of the system self-heating (the change in the temperature of the material due to the chemical reaction: positive for an exothermic reaction and negative for an endothermic reaction), which directly influences the reaction rate. Thus, the effects of heat release during the reaction as well as heat exchange parameters (heat-transfer coefficient, heating rate, thermal effect of the reaction, sample weight and others) cannot be taken into consideration. To improve the situation, a sample weight and a heating rate (and some other parameters) corresponding to the absence of self-heating are sometimes assigned a priori using purely thermophysical approaches without taking into account the heat from the chemical reaction. A quasi-stationary temperature lag ($\Delta T_{qs} = \omega r^2/a$, where ω is the linear heating rate, r is the sample radius, and a is the sample temperature diffusivity) and the conditions corresponding to a small ΔT_{qs} are selected.

   Such evaluations are possible only for a limited parameter range; in particular for low thermal effects which do not result in noticeable self-heating. In the general case, self-heating temperatures can exceed the quasi-stationary temperature difference by many times and even by orders of magnitude [8].

The question of why one needs to know the heat-exchange conditions if they are not used in the experimental technique for determining the reaction rate (for example, in thermogravimetry and scanning calorimetry) often arises. In this case, thc heat exchange parameters are not needed for the direct measurement or calculation of the reaction rate. However, information on the heat-exchange conditions is required to identify the temperature that the determined reaction rate is attributed to. *In this case, to achieve the correct quantitative analysis of the experimental data, one should ensure that a uniform temperature distribution exists over the reaction volume and measure the sample temperature.*

In the general case, the complete elimination of self-heating is not necessary. However, it is required if the temperature spatial distribution is not uniform. Methods for equalizing the temperature and decreasing the self-heating are considered below.

b) The second group includes approaches based on the joint analysis of kinetics and heat balance equations written under the assumption of a uniform temperature distribution over the reaction volume (see for example [7, 11, 12]). This description of the studied system is closer to reality and more accurate than the first one, provided that a uniform temperature distribution is ensured in the experiment. The parameter ranges corresponding to a uniform temperature distribution cannot be evaluated using this analytical approach. However, comparison with a more general case enables one to conclude that it is quite wide. Therefore, the heat balance equation combined with the kinetic equation is used as the basis for formulating the reverse kinetic problem.
c) In the third group of approaches, the kinetic equation is analyzed jointly with the thermal conductivity equation, which ensures a more accurate and complete description of the studied process. In this case, the computation is quite complicated, even when diffusion and hydrodynamic factors are neglected. The main assumption used for this group of approaches is that the degree of conversion is independent of the spatial coordinates. The use of this assumption results in a significant smoothing of the temperature distribution, which can influence how the process characteristics depend on the corresponding parameters.

In conclusion, let us consider some results from theoretical studies in which the spatial distribution of the temperature and the degree of conversion were taken into consideration [13, 14]. After the solving for the distributions of the temperature (T) and the conversion degree (η) in a sample undergoing linear heating, it was shown that

i) For most practically interesting materials (including explosives and blasting powders), the critical rates are quite low, even for small samples. In other words, small samples inflame at almost all of the heating rates used in standard DTA and TGA devices and scanning calorimeters ($\omega \approx 1$–$100°C/min$).
ii) At standard conditions for thermogravimetric experiments, the degree of nonuniformity of the temperature distribution in a sample, evaluated using the sample weight on the basis of the "nonstationary" temperature difference, is significantly underestimated and as a rule should not be used in kinetic analysis.

In [13, 14], the complete set of equations describing the macrokinetics of exothermic and endothermic decomposition upon linear heating was considered and the fundamental role of the degeneration parameter used in the thermal explosion theory (the Todes number)

$$\gamma = \frac{c\rho}{Q}\frac{RT_*^2}{E},$$

where T_* is the critical temperature of the thermal explosion under stationary conditions (that is, $T = \mathrm{const}$), was quantitatively analyzed.

Some other results from [13] and [14] relating to thermographic studies of systems with high values of the degeneration parameter γ are discussed in the following sections.

6.2 Methods of Equalizing the Temperature and Decreasing Self-Heating: Mechanical and Thermal Dilution

Let us consider methods that ensure the correctness of the experimental data on reaction kinetics obtained under programmed heating. As was mentioned above, commercially available devices are not designed to study reaction kinetics. Therefore, in the general case, one cannot clearly identify the heat exchange conditions between the environment and the experimental cell based on its design. This complicates the quantitative analysis of the experimental data and markedly decreases the accuracy of the estimated kinetic parameters. Special approaches that are intended to improve this situation are proposed in [7, 11, 15, 16, 17]. For example, in an experimental study of the kinetics of reaction in solution [11], the liquid reacting system was vigorously stirred. This allowed the process to be described by a set of equations in which averaged characteristics were used. For obvious reasons, this method cannot be applied to solid materials, including components of composite solid propellants.

Another approach based on the use of an experimental cell characterized by very weak heat dissipation into the environment is proposed in [7]. In this case, the temperature distribution over the reaction volume is quite uniform for liquid and solid systems. However, the use of this method does not exclude thermal explosion for highly exothermal reactions and, hence, it does not widen the extremely narrow range of measurable reaction rates.

A method of equalizing the temperature based on the use of very light samples (several μg) cannot be considered to be universal. It cannot be applied to study the kinetics of heterogeneous systems containing coarse particles (fractions of mm), including composite solid propellants. However for the study of "liquid–fine solid" systems that react with low heat release, a combination of the methods proposed in [7] and [11] may be quite promising.

In [15, 16, 17] we proposed an approach that not only ensured low coefficients of heat exchange between the sample and the thermostat, but that also equalized the temperature and at the same time decreased the self-heating via significant ballasting ("dilution") of the reactant with an inert compound (the "diluent") with a high thermal conductivity. This approach is characterized by a number of advantages.

Firstly, the possibility of thermal explosion for exothermal reactions is completely eliminated (a transfer to completely degenerated modes takes place due to increasing γ). This allows one to study processes with higher thermal effect at higher temperatures and to use standard thermal analysis heating rates.

Secondly, the calculations result in relatively accurate results due to the almost complete elimination of temperature gradients.

Thirdly, the thermophysical characteristics of the studied material are not required, since the heating conditions are determined by the properties of the diluent. In nonisothermal kinetic experiments, the change in the sample composition cannot be described analytically, even if the expression describing the reaction kinetics is available. Therefore, the thermophysical characteristics of the sample material cannot be taken into account quantitatively. This circumstance is an additional argument in favor of the methods developed by us. Besides, a reference substance for use in DTA is not required, since the role of this substance is played by the diluent.

Two methods of dilution were developed: "mechanical" and "thermal." In the first method, the reagent is blended with powder of the heat-conducting chemically inert diluent and placed into a vial. In the second one, a thin (<1 mm) layer of reagent is placed between heat-conducting blocks (Fig. 6.1).

The first method is applied for the study of homogeneous solid and liquid materials, while the second one is used for homogeneous and heterogeneous condensed materials. In both cases the diluent is used as a reference substance. The size and shape of the reference cell are similar to those of the working cell. A typical dilution ratio used is 1/100 (which does not affect the accuracy of TGA data recorded in the experiments using analytical adjusted beam scales).

The diluent should not exhibit any catalytic behavior in the studied reaction. This condition is especially important in case of mechanical dilution, since the reagent–diluent contact area is very large. In the case of thermal dilution, the contact area is relatively small, and the catalytic effect of the heat-conducting blocks can be neglected. In some cases, to completely eliminate the influence of the block material, thin layers of inert insulator (for example, mica) can be placed between the sample and the blocks. Obviously, a high degree of ballasting results in output signal depression. However, this problem can be resolved by using commercially available photovoltaic amplifiers.

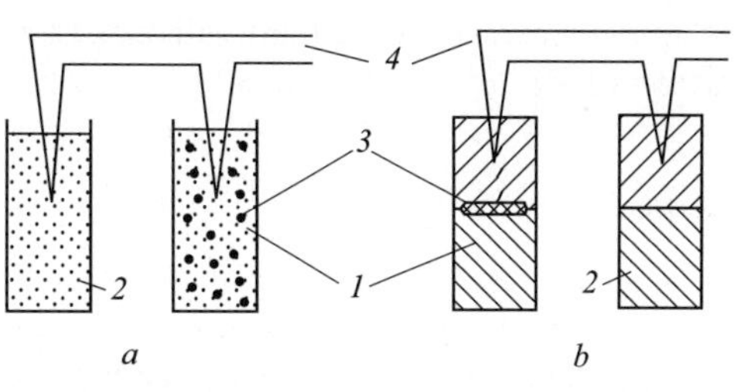

Fig. 6.1a–b Schematic illustrating "mechanical" (**a**) and "thermal" (**b**) dilution of a reagent with an inert compound in thermography; *1*, working cells; *2*, reference cells; *3*, reagent; *4*, differential thermocouples

6.3 Estimation of Kinetic Constants from DTA–TGA Data Obtained by Dilution-Based Techniques

Let us consider the heat balance for the thermograph cells shown schematically in Fig. 6.1.

For the working cell

$$(CM+cm)\frac{d(T_{\text{ref}}+\Delta T)}{dt}=Qm\frac{d\eta}{dt}-\alpha S\left[(T_{\text{ref}}+\Delta T)-T_0\right]. \tag{6.1}$$

For the reference cell

$$CM\frac{dT_{\text{ref}}}{dt}=-\alpha S(T_{\text{ref}}-T_0), \tag{6.2}$$

with the initial conditions $t=0$, $T_{\text{ref}}=T_0=T_{0\text{in}}$, $\Delta T=0$, $\eta=0$, where $T_{0\text{in}}$ is the initial temperature of the oven wall, T_{ref} is the reference substance temperature, C and M are the heat capacity and weight of the reference substance, while c and m are the heat capacity and weight of the reagent.

Subtracting Eq. (6.2) from Eq. (6.1) and using inequality $CM \ll cm$, one obtains the following original equations for the DTA signal:

$$CM\frac{d(\Delta T)}{dt}=Qm\frac{d\eta}{dt}-\alpha S\Delta T, \tag{6.3}$$

$$\frac{d\eta}{dt}=\dot{\eta}(T,\eta), \tag{6.4}$$

with the initial conditions $t=0$, $\Delta T=0$, $\eta=0$.

Equation (6.3) is strictly valid for a static (isothermal) mode. In this case, the temperature dependence of the thermophysical parameters of the cell can be neglected under the assumption that $C=\text{const}$ and $\alpha=\text{const}$ in the experiment. Under the conditions of programmed heating (dynamic modes), instead of Eq. (6.3) one can write

$$C(T)M\frac{d(\Delta T)}{dt}=Qm\frac{d\eta}{dt}-\alpha(T)S\Delta T \tag{6.5}$$

with the same initial conditions.

Integration of the left hand side of Eq. (6.5) with respect to t over the interval 0–t_{end} (the moment at which the process ends, when $\eta\approx 1$) using the tabular temperature dependence of the inert substance (intermediate mathematical manipulations are omitted)

$$C(T)=a+bT, \tag{6.6}$$

gives, for dynamic heating with rate $T'=\omega$,

$$M\int_0^t C(T)\frac{d(\Delta T)}{dt}dt=M\omega b\int_0^{t_{\text{end}}}\Delta T\,dt. \tag{6.7}$$

The corresponding expression for the thermal effect is

$$Q = \frac{1}{m} \int_{0}^{t_{\text{end}}} [\alpha(\mathrm{T})\mathrm{S} - \mathrm{M}\omega b] \Delta T dt. \tag{6.8}$$

Expressions for the reaction rate and degree of conversion (omitted for the sake of brevity) also contain a correction term $M\omega b$ that can be neglected in practice. Since for commonly used diluents (Al, Al_2O_3, SiO_2 and others), the value of parameter b in Eq. (6.6) is lower than or equal to $4 \times 10^{-3}\,\mathrm{J\,deg^{-2}\,g^{-1}}$, the following inequality is always valid for moderate heating rates ($\omega \leq 10\,\mathrm{deg\,min^{-1}}$):

$$\frac{M\omega b}{\alpha(T)S} << 1.$$

Thus, in contrast to the temperature dependence $\alpha = \alpha(T)$ (which is determined in a special experiment as $S\alpha(T)$, where S is the heat-exchange surface; see Chap. 7), $C = C(T)$ can be neglected in the calculation of Q, η and $\dot{\eta}$. In this case, the diluent heat capacity $C(\overline{T})$ corresponding to the average experimental temperature $\overline{T}$ can be substituted into the formulae.

Thus, to obtain an intermediate data array including the thermal effect (Q) and sets of values for the temperature (T), conversion degree (η) and reaction rate ($\dot{\eta}$) for a certain time (t), one should measure the time dependence of the DTA signal $\Delta T = \Delta T(\omega,t)$ and that of the reference substance temperature $T_{\text{ref}} = T_{\text{ref}}(\omega,t)$.

After differentiating $\Delta T = \Delta T(\omega,t)$ with respect to time, Q and $\dot{\eta} = \dot{\eta}(T,\eta)$ can be obtained via the following set of equations derived from Eqs. (6.3) and (6.4):

$$Q = \frac{1}{m} \int_{0}^{\infty} \alpha(T) S \Delta T dt, \tag{6.9}$$

$$\eta = \eta(\omega,t) = \frac{1}{Qm} \left[MC\Delta T + \int_{0}^{t} \alpha(T) S \Delta T dt \right], \tag{6.10}$$

$$\dot{\eta} = \dot{\eta}(\omega,t) = \frac{1}{Qm} \left[MC(\Delta\dot{T}) + \alpha(T) S \Delta T \right]. \tag{6.11}$$

Since the relationship $T = T_{\text{ref}} + \Delta T = T(\omega,t)$ is known, Eqs. (6.10) and (6.11) can be considered a parametric form of the kinetic equation. Indeed, the dependence $\dot{\eta} = \dot{\eta}(T,\eta)$ can be obtained by eliminating ω and t from them.

Let us graphically illustrate the described analytical approach. The experimental DTA curves ($\Delta T = \Delta T(\omega,t)$) obtained for various linear heating rates (ω_1, ω_2, ω_i) and corresponding plots of the working cell temperature against time ($T = T(\omega,t)$) are shown in Fig. 6.2a,b. Analysis of these data using Eqs. (6.10) and (6.11) yields the time dependencies of the conversion degree ($\eta = \eta(\omega,t)$) and the reaction rate ($\dot{\eta} = \dot{\eta}(\omega,t)$) for the same linear heating rates (Fig. 6.2c, d). Drawing a straight

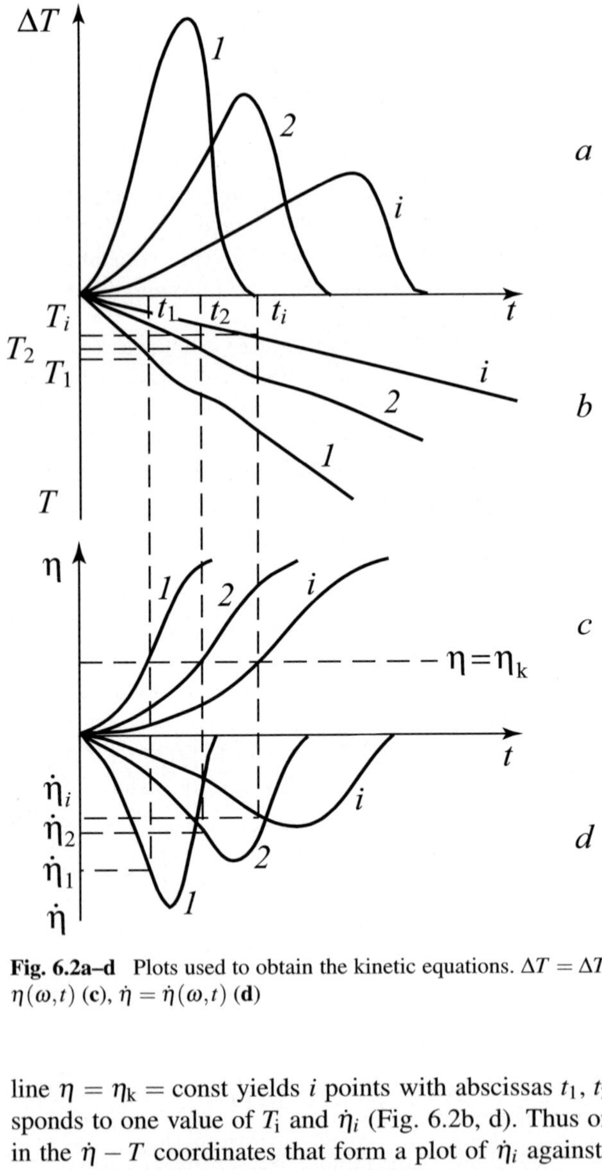

Fig. 6.2a–d Plots used to obtain the kinetic equations. $\Delta T = \Delta T(\omega,t)$ (**a**), $T = T(\omega,t)$ (**b**), $\eta = \eta(\omega,t)$ (**c**), $\dot{\eta} = \dot{\eta}(\omega,t)$ (**d**)

line $\eta = \eta_k =$ const yields i points with abscissas t_1, t_2, t_i. Each abscissa t_i corresponds to one value of T_i and $\dot{\eta}_i$ (Fig. 6.2b, d). Thus one obtains a locus of points in the $\dot{\eta} - T$ coordinates that form a plot of $\dot{\eta}_i$ against T. Since each value of η_1, η_2, η_k corresponds to the curve $\dot{\eta}(T)$, one obtains a set of k curves $\dot{\eta} = \dot{\eta}(T)$ at $\eta =$ const representing the graphical form of $\dot{\eta} = \dot{\eta}(T,\eta)$ (Fig. 6.3a). The number of experimental points on the $\dot{\eta}(T)$ curve at $\eta =$ const equals the number of experiments i for various heating rates ω_i. For processes that comprise one stage overall, data (Fig. 6.3a) can be presented as an Arrhenius anamorphosis in $\ln \dot{\eta} - T^{-1}(K^{-1})$ coordinates at $\eta =$ const (Fig. 6.3b). The temperature interval for the dependence $\dot{\eta} = \dot{\eta}(T,\eta)$ can be widened by extrapolation (dotted lines in Fig. 6.3b).

Fig. 6.3 Plots used to obtain kinetic equations. $\dot{\eta} = \dot{\eta}(T)$ at $\eta = \text{const}$ (**a**, **b**), $\dot{\eta} = \dot{\eta}(\eta)$ at $T = \text{const}$ (**c**)

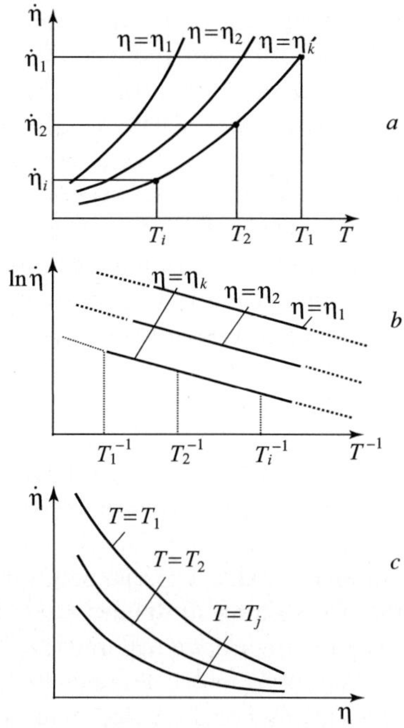

The set of curves presented in Fig. 6.3a, b can also be obtained using a different method. By drawing horizontal lines at $T = T_\mathrm{j} = \text{const}$ in Fig. 6.2b, one can determine η and $\dot{\eta}$ at corresponding time t_j and plot j isothermal curves of $\dot{\eta} = \dot{\eta}(\eta)$ at $T = \text{const}$ (Fig. 6.3c). The number of points on these curves is also equal to the number of experiments i at various heating rates ω_i. From experience of numerous computations, the first method of determining the function $\dot{\eta} = \dot{\eta}(T,\eta)$ is more convenient, while the second technique allows one to obtain the $\varphi(\eta)$ function directly.

To further analyze the $\dot{\eta}(T)$ dependencies at $\eta = \text{const}$ or the $\dot{\eta} = \dot{\eta}(\eta)$ dependencies at $T = \text{const}$, standard chemical kinetics methods are used (an equation describing these dependencies is selected, the parameters of the equation are determined; the kinetic equation obtained corresponds to known reaction types or it is purely empirical).

The proposed method of nonisothermal data recalculation for isothermal conditions in TGA experiments is significantly simpler. From the experimental data on $T = T(\omega,t)$ (the temperature measured in the working cell) at various ω_i (Fig. 6.2b, c) and the corresponding dependencies $\dot{\eta} = \dot{\eta}(\omega,t)$ [the time derivatives of dependencies $\eta = \eta(\omega,t)$, Fig. 6.2d], one can easily obtain the dependencies $\dot{\eta} = \dot{\eta}(T,\eta)$ (Fig. 6.3a) using the described method, etc.

To conclude, let us consider some aspects of the applicability of the proposed experimental and analytical methods. The experimental data can be considered correct if $t_\mathrm{r} \gg t_\mathrm{h}$, where t_r is the characteristic reaction time and t_h is the cell thermal relaxation time. In the first approximation, t_h can be assumed to be equal to the setting time for quasi-stationary heating t_{quasi} (if the DTA curve is approximated by a set of line sections corresponding to various heating/cooling rates). For typical cells and samples (see Chap. 7), one finds by using expressions for nonstationary heat transfer [18] that $t_{quasi} \approx 0.8$–1 s. In other words, the method is valid for studying reactions characterized by reaction times of 10 s or longer.

To exclude the self-inflammation of a sample undergoing linear heating, the reagent should be significantly diluted with an inert compound $\left(\frac{cm}{CM} \ll 10^{-2}\right)$. However, sample self-inflammation can happen if the particle diameter d (in the case of mechanical dilution) or the layer thickness l (in the case of thermal dilution) is not small enough. According to estimates obtained using methods related to the dynamic theory of thermal explosion [8] (Table 6.1), the critical heating rates for self-inflammation of dinitroxydiethylnitramine (DINA) and nitrocellulose noticeably exceed the maximum heating rates used during DTA with $d \leq 2 \times 10^{-2}$ cm and $l \leq 3 \times 10^{-2}$ cm. The mechanical dilution method was thoroughly studied by Barzykin [21], who showed that for a hundredfold dilution of nitrocellulose with aluminum oxide, the maximum temperature difference between the central part of the cell (0.6 cm in diameter) and its wall is still smaller than the quasistationary temperature drop (equal to 0.09°C) at $\omega = 20\,\mathrm{deg\,min^{-1}}$.

The applicability of nonisothermal methods to the study of complex (multistage) reactions is not discussed here (some additional information on this topic can be found in [5]). The analytical approach described in this chapter can be used to study a wide range of gross one-stage processes described by one kinetic equation (for example, simple n-order, reversible and self-catalyzed reactions). It can also be useful

Table 6.1 Critical rates of linear heating for mechanical and thermal dilution

Compound	Kinetic equation [reference]	Dilution type: Mechanical $d = 0.02$ cm Al_2O_3		Dilution type: Thermal $l = 0.03$ cm Al	
		Critical rates, °C min^{-1}			
		ω^*_{qs}	ω^*_{pi}	ω^*_{qs}	ω^*_{pi}
DINA	$\dot{\eta} = 4.3 \times 10^{18} \exp\left(-\frac{45000}{RT}\right)(1-\eta)$ [19]	300	109	64	24
Nitrocellulose	$\dot{\eta} = 10^{19} \exp\left(-\frac{47000}{RT}\right)(1-\eta) + 10^{12} \exp\left(-\frac{31000}{RT}\right)\eta(1-\eta)$ [20]	310	113	157	58

Here ω^*_{qs} and ω^*_{pi} are the critical heating rates under the quasistationary and pseudoisothermal approximations, respectively (the values of the critical rates shown here are approximate and limiting, $\omega^*_{qs} > \omega^* > \omega^*_{pi}$, where ω^* is the true critical rate value).

when the reaction stages are clearly separated (the DTA peaks are clearly spaced) and can be treated as independent reactions.

References

1. Merzhanov AG (1973) Combust Explos Shock Waves 9:4
2. Kozlov YuN, Nadezhdin AD, Pourmal AP (1974) Int J Chem Kinet 6:383
3. Berg LG (1969) Basics of thermography. Nauka, Moscow
4. Piloyan GO (1964) Basics of thermal analysis theory. Nauka, Moscow
5. Merzhanov AG, Barzykin VV, Shteinberg AS, Gontkovskaya VT (1977) Thermochim Acta 21:103
6. Sestak J (1984) Thermophysical properties of solids. Academia, Prague
7. Merzhanov AG, Abramov VG, Abramova LT (1967) Rus J Phys Chem 41:90
8. Barzykin VV (1973) Combust Explos Shock Waves 9:29
9. Murray P, White J (1955) Trans Brit Ceram Soc 54:4
10. Kissenger HE (1957) Anal Chem 29:1702
11. Borchardt HJ, Daniels F (1957) J Am Chem Soc 79:41
12. Reed PL, Weber L, Gottfried BS (1965) Ind Eng Chem Res 41:39
13. Gontkovskaya VT, Barzykin VV (1974) Izv Sib Otd Akad Nauk SSSR Ser Khim Nauk 4:68
14. Barzykin VV, Gontkovskaya VT (1974) The effect of temperature distribution on the study of chemical reactions kinetics under programmed heating. Institute of Chemical Physics of the USSR Academy of Sciences, Chernogolovka, Russia
15. Goncharov EP, Driatskaya GI, Merzhanov AG, Shteinberg AS (1971) Dokl Chem 197:254
16. Goncharov EP, Merzhanov AG, Shteinberg AS (1972) In: Stesik LN (ed) Combustion and explosion. Nauka, Moscow, p 765
17. Goncharov EP, Merzhanov AG, Shteinberg AS, Ledovskaya NA (1973) In: Tyuryaev IYa (ed) Catalysts of basic organic syntheses. Khimiya, St. Petersburg, p 60
18. Kustov YuA, Rybanin SS (1970) Combust Explos Shock Waves 6:50
19. Dubovitksy FI, Rubtsov YuI, Barzykin VV, Manelis GB (1960) Izv Akad Nauk SSSR Ser Khim 6:1126
20. Manelis GB, Rubtsov YuI, Smirnov LP, Dubovitsky FI (1962) Kinet Catal 1:32
21. Barzykin VV (1977) Thermal analysis of reacting compounds. Institute of Chemical Physics of the USSR Academy of Sciences, Chernogolovka

Chapter 7
Equipment for Studying Kinetics During Linear Heating

Abstract Several DTA and TGA instruments used to study the kinetics of fast high-temperature reactions in condensed phases of energetic materials over a wide range of temperatures and pressures are described. The design of new experimental cells used in original approaches to the "thermal and mechanical dilution" of the sample with inert heat-conducting material is discussed in detail. Due to the special cell design, the sample is spatially isothermal and even very fast exothermal reactions do not result in self-inflammation. The use of the special cell in combination with commercial DTA, TGA and DSC instruments allows one to increase the upper limit of measurable reaction rates by several orders of magnitude.

7.1 Thermographic Instruments

Two types of setups have been used in DTA experiments: the special thermograph T-1 (Fig. 7.1) and the universal thermographs T-2 and T-3 (Figs. 7.2 and 7.3).

Thermograph T-1 can be used in experiments at atmospheric pressure and in vacuum ($10^5\,\text{Pa} \geq P \geq 10\,\text{Pa}$), while the universal thermographs T-2 and T-3 can also be used at elevated pressures ($3\,\text{MPa} \geq P \geq 10\,\text{Pa}$). The thermographic setups consist of a thermograph, a recording system, a pressurizing system, a temperature programming unit (TPU), a set of working cells, and instruments for sample preparation (sieves, grinding mills, cutters, etc.).

7.1.1 Thermographs

The main unit of thermograph T-1 (Fig. 7.1) is an aluminum cylinder 70 mm in diameter, 180 mm long (*5*) with two symmetrical holes for cells and working thermocouples. Heater *6* and the reference chromel-alumel thermocouple *7* (the junction of which is separated from the heater by a thin layer of mica) are fixed onto the lateral surface of the cylinder. To diminish outward heat losses, the heater is covered with a glass cloth and placed into a ceramic casing. Chromel-alumel working

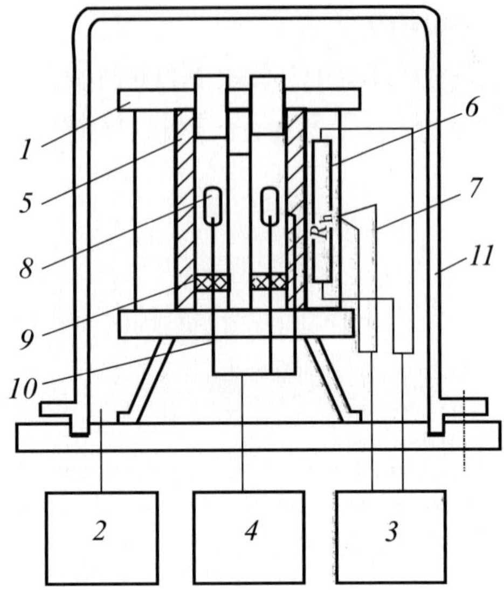

Fig. 7.1 Schematic diagram of thermographic setup T-1: *1*, thermograph; *2*, pressurizing system; *3*, temperature programmer; *4*, registration unit; *5*, thermograph block (Al cylinder); *6*, heater; *7*, reference thermocouple; *8*, working cell; *9*, centering ring plate; *10*, working thermocouple; *11*, protective casing

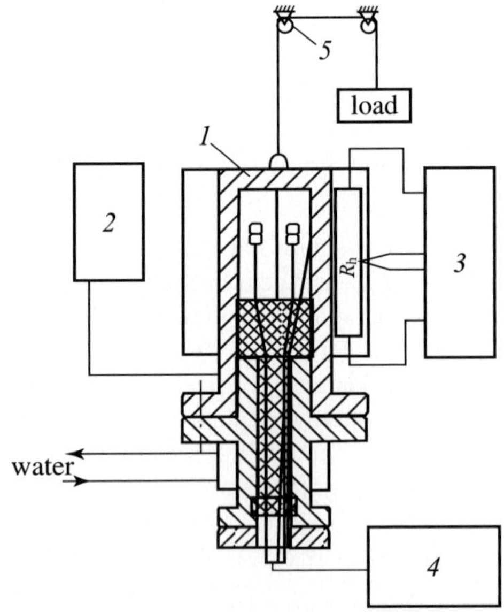

Fig. 7.2 Schematic diagram of thermographic setup T-2: *1*, thermograph; *2*, pressurizing system; *3*, temperature programmer; *4*, registration unit; *5*, system for casing manipulation

thermocouples *10* are inserted into the holes in the block and centered in the cylinder via ceramic and aluminum ring plates *9*. These act as thermal barriers, preventing heat losses from the cells via the thermocouple wires. This is extremely important in DTA with dilution, since the heat released in a working cell is not very high.

The working cells *8* are mounted on the thermocouple junction. The ends of the thermocouples are placed into thin (1 mm in diameter) porcelain two-channel tubes, while the thermocouples themselves are placed in thicker (3 mm in diameter) tubes.

Fig. 7.3 Photo of thermographic installation T-3

The working thermograph T-2 unit (Fig. 7.2) is placed into a thick-walled cylindrical stainless steel split casing used as a heater. A heating coil and a control thermocouple are attached to the external surface of the upper part of the casing, which can move vertically, providing easy access to the cells. The thermocouples are centered using ring plates (thermal barriers). The thermocouple outputs are sealed with Teflon powder. The upper part of the casing is tightly attached to the lower part with three link bolts. The sealing interface is cooled with flowing water.

7.1.2 Pressurizing System

The required working atmosphere is provided by a pressurizing system consisting of a cylinder with inert gas, a vacuum pump, a receiver, valves, and regular and vacuum manometers.

7.1.3 Experimental Procedure

In the experiments, the following parameters are continuously recorded by a computer or a multiple-point potentiometer: the DTA signal, the working cell temperature and the heater (cylinder) wall temperature. The DTA signal was amplified by a microvoltmeter/microammeter (*1*) connected to a computer and an electronic potentiometer via a resistance box (*2*) (Fig. 7.4). The scale interval of the recorder was in the range of $(0.75\text{–}1.5)\times10^{-2}$ deg mm^{-1}, depending on the preset resistance.

Potentiometers equipped with constant-voltage power supply units were connected to the circuit to measure the heater and working cell temperatures. The potentiometers (*3, 4*) were connected to the thermocouples, which ensured that highly accurate temperature measurements were obtained (the recorder scale interval was equal to 0.3 deg mm^{-1}). In thermograph T-3 this procedure was automated using an automatic potentiometer designed on the basis of a step-by-step selector.

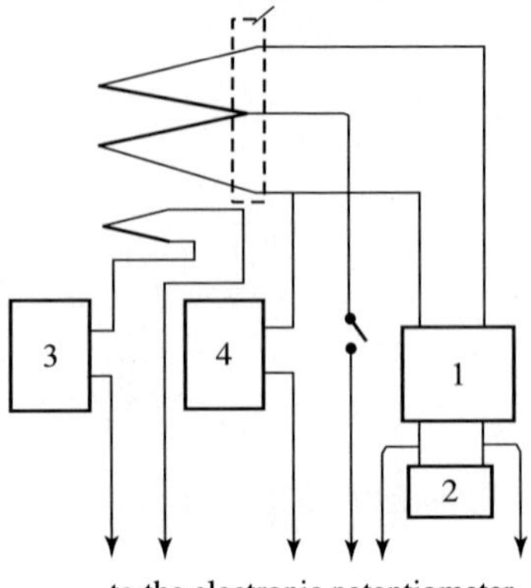

Fig. 7.4 Scheme of the data recording in DTA: *1*, microvoltmeter/microammeter; *2*, resistance box; *3* and *4*, potentiometers

7.1.4 Working Cells

The working (reaction) cells used in the DTA experiments with thermal dilution and the TGA experiments are almost identical; see Fig. 7.5 (*1, 2*).

The working cell consists of a thin-walled (0.2–0.4 mm) aluminum or stainless steel cylinder 7 mm in diameter with a hole or a curling in the bottom (*a*) and two aluminum cylinders (*b*). The sample, in the form of a thin (0.2–0.4 mm) layer of powder (homogeneous compounds, model mixtures) or a thin disk (composite solid propellants), is placed between the strictly parallel ends of the aluminum cylinders. Tight sample–cylinder contacts are achieved by using a forcing screw or a bayonet lock. Gaseous decomposition products are removed through a number of channels (0.6 mm in diameter) in the cylinders.

The reaction cell used in the experiments with mechanical dilution (see Fig. 7.5 (*3*)) is similar to a standard thermographic cell.

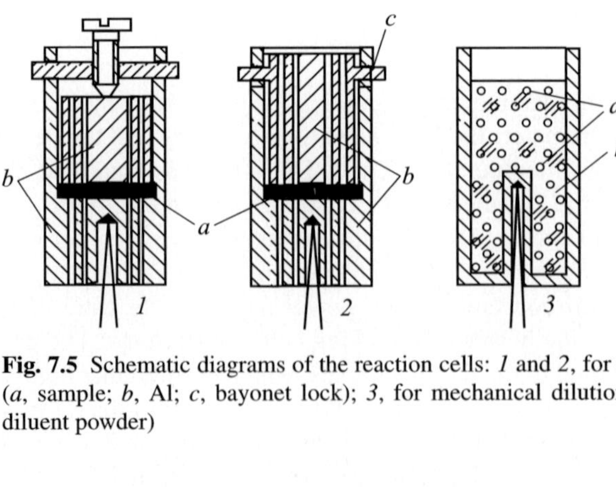

Fig. 7.5 Schematic diagrams of the reaction cells: *1* and *2*, for thermal dilution in DTA and TGA (*a*, sample; *b*, Al; *c*, bayonet lock); *3*, for mechanical dilution in DTA (*a*, sample particles; *b*, diluent powder)

7.2 Thermogravimetric Instrument

The thermogravimetric setup was based on computer-assisted high-accuracy (2 mg scale range) thermogravimetric scales (thermo-scales) [1] combined with a system for cell (sample) temperature registration and a temperature-programming device (Fig. 7.6). Due to the weight change compensation system used in the thermo-scales (a balancing solenoid affects a magnet that is rigidly connected to the sample and the weight beam), the position of the cell (suspended on a thread) is constant. Therefore, measuring the cell temperature using a thermocouple does not affect the recording of changes in weight. The temperature measurement procedure is similar to the one described in Sect. 7.1.3.

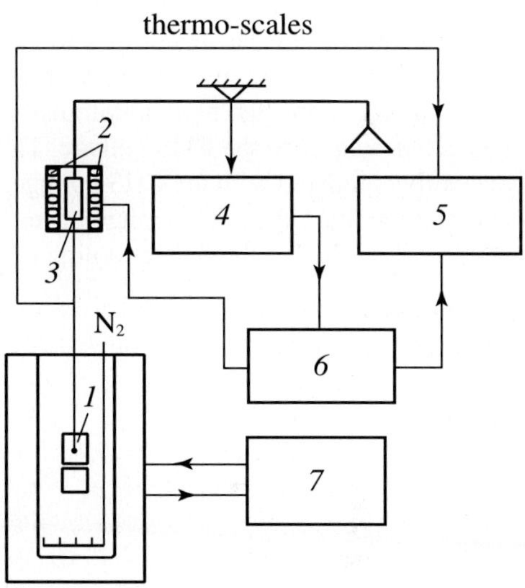

Fig. 7.6 Scheme of thermogravimetric setup: *1*, sample (working cell); *2*, solenoid coil; *3*, solenoid core; *4*, photosensor; *5*, registration unit; *6*, control unit; *7*, temperature programmer

7.3 Temperature Programmer

Linear heating of a sample can be performed experimentally by various means. The simplest method is based on changing the temperature linearly in a liquid thermostat at a constant contact thermometer magnetic head rotation rate [2]. However this method can only be applied over a narrow temperature range. A number of approaches that provide linear heating across a wide range of temperatures with high accuracy are described in [3, 4, 5]. In [3], the temperature difference between the internal and external surfaces of the oven wall was kept constant during the heating. In this case, one junction of a differential thermocouple is located on the internal wall, while the other one is placed near the heater. The heating rate can be changed by controlling the signal from the differential thermocouple by setting a contact galvanometer limiter or by moving dial galvanometer photocells. The heater power is controlled in the relay mode by using two photorelays, or smoothly by using an autotransformer connected to a reversible electric motor. The methods described in [4, 5] are based on a null control principle (astatic system). Voltage variations in a Pt thermistor corresponding to the changes in the thermostat temperature are compared (using the galvanometers and photorelays) with the scheduled voltage on the heating rate sensor. In [5], a thyristor-based circuit was used instead of standard photorelays.

In our experiments, a temperature programmer for linear heating (a follow-up system with proportional-plus-floating control and an electronic heating rate sensor) designed based on previous experience was used (Fig. 7.7, Table 7.1).

The computer-assisted heating rate sensor (CHS) generates a varying voltage which corresponds to a change in the chromel-alumel thermocouple voltage during linear temperature growth at a preset rate, ω. This voltage is added to the thermoelectric voltage of the reference thermocouple (of the opposite sign) installed near the thermostat heater. The resulting voltage is received by an amplifier (null device). If an imbalance occurs, the current in the heater circuit is adjusted by a reversible electric motor connected to an autotransformer cursor until the thermoelectric voltage becomes equal to the CHS voltage. Thus the reference thermocouple voltage is constantly compared with the CHS voltage, which results in the linear growth of the heater temperature for a wide range of temperatures and heating rates (this method can also be applied to the linear cooling of a sample).

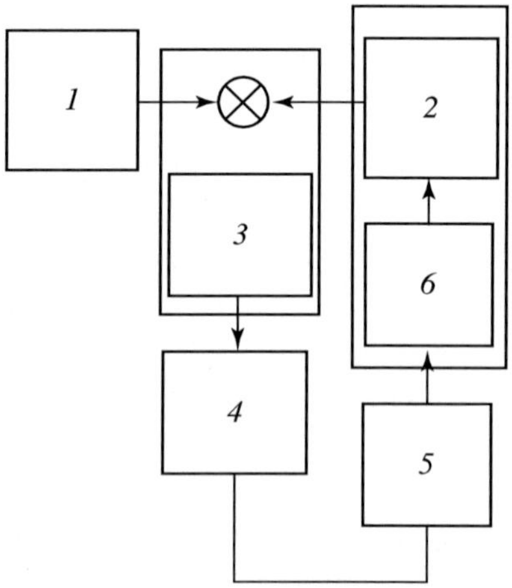

Fig. 7.7 Scheme of temperature programmer: *1*, computer-assisted heating rate sensor (CHS); *2*, control thermocouple; *3*, dc amplifier; *4*, reversible motor; *5*, autotransformer; *6*, heater

Table 7.1 Technical characteristics of the temperature programmer

Working temperature	40–600°C
Heating rate	0.1–20 deg min^{-1}
Maximum heating nonlinearity for chromel-alumel	±1%
Maximum heating power	1 kW
Supply voltage	220 V

References

1. Ga'perin LN (1962) Zh Fiz Khim 36:2294
2. Shelyubsky VI (1959) Prib tekh Exper 5:84
3. Berg LG (1969) Basics of thermography. Nauka, Moscow
4. Gal'perin LN (1963) Inzh-Fiz Zh 1:39
5. Kolesov YuR (1970) PhD thesis. Institute of Chemical Physics of the USSR Academy of Sciences, Chernogolovka, Russia

Chapter 8
Application of Modified DTA and TGA to Studies of High-Temperature Decomposition Kinetics of Homogeneous Compounds

Abstract Some results from studying the kinetics of fast high-temperature exothermal reactions in homogeneous energetic materials (nitrocellulose, ammonia copper chromate and ammonium perchlorate (AP)) using DTA and TGA are presented. In the experiments, specially designed cells (described in Chap. 7) were used. In these cells, the sample is spatially isothermal and even very fast exothermal reactions cannot cause self-inflammation. Maximum temperatures corresponding to the measured reaction rates and kinetic constants are close to those in the reaction zone near the burning surface. Taking the decomposition of nitrocellulose as an example, a stage characterized by high activation energy is shown to dominate at high temperatures, while at low temperatures its role is negligible. The kinetic mechanism of the high-temperature decomposition of AP is shown for the first time to be identical to that for low-temperature decomposition. However, the macrokinetics of the high-temperature decomposition of AP is found to differ qualitatively from that of the low-temperature process. This can be explained by the occurrence at high temperatures of dissociative sublimation at thermodynamic equilibrium. This effect should be taken into account when analyzing the mechanisms of high-temperature processes such as the ignition and burning of AP-based solid rocket propellants.
The application of thermal and mechanical dilution methods to studies of the decomposition kinetics of homogeneous (individual) compounds is illustrated for nitrocellulose (a typical blasting powder), ammonia copper chromate (which decomposes with the formation of copper chromite, an efficient combustion catalyst, and a component in a number of commercial solid propellants) and AP (the main oxidizer used in modern composite solid propellants).

8.1 High-Temperature Decomposition of Nitrocellulose

Experimental and calculated data on the high-temperature decomposition of nitrocellulose obtained using TGA are presented in Figs. 8.1–8.4. The experiments were carried out in air at atmospheric pressure using installation T-1. For the mechanical dilution of nitrocellulose ($MC/mc \approx 100$), burnt Al_2O_3 powder was used (the use of

Fig. 8.1 Thermograms of the high-temperature decomposition of nitrocellulose at heating rate $\omega = 12$ (*1*), 9.5 (*2*), 6.2 (*3*), 4.9 (*4*), 2.8 (*5*) deg min^{-1}

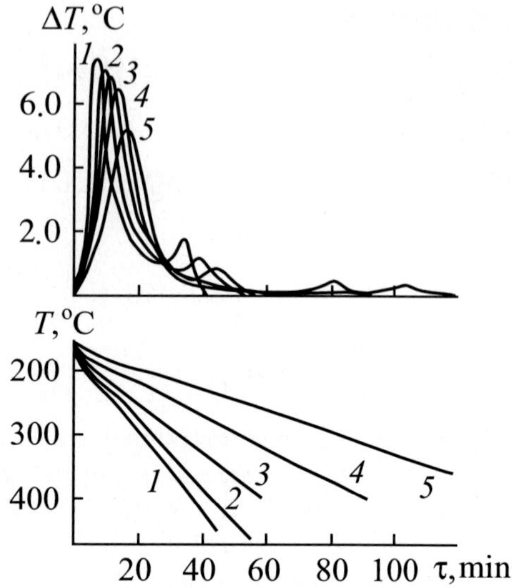

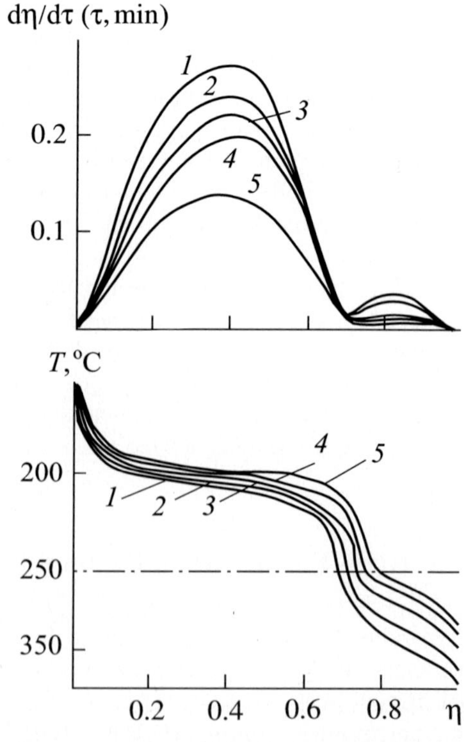

Fig. 8.2 Plots of the non-isothermal rate of nitrocellulose decomposition against the conversion degree at $\omega = 12$ (*1*), 9.5 (*2*), 6.2 (*3*), 4.9 (*4*), 2.8 (*5*) deg min^{-1}

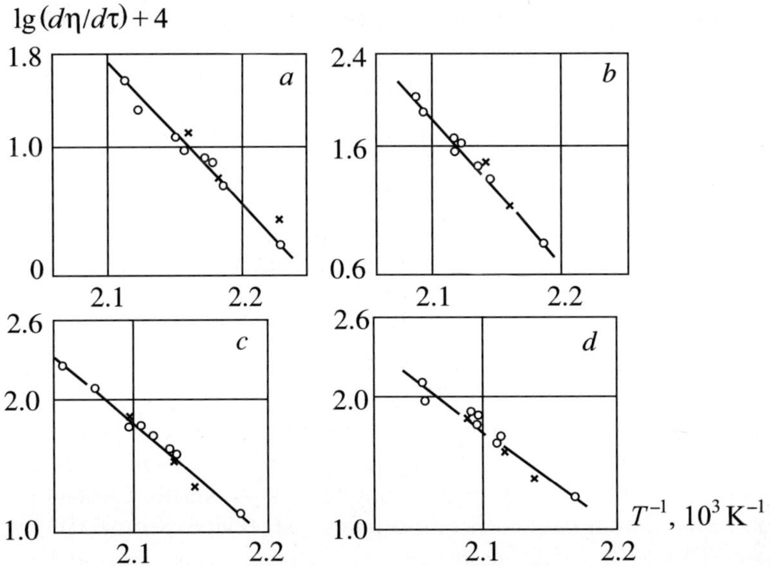

Fig. 8.3 Plots of the nonisothermal rate of nitrocellulose decomposition against the temperature at $\eta = 0.1$ (**a**), 0.2 (**b**), 0.4 (**c**), 0.6 (**d**) for dilution with Al_2O_3 (*circles*) and SiO_2 (*crosses*)

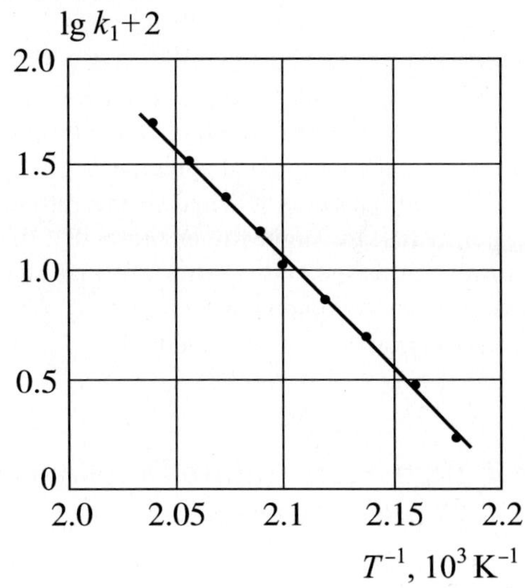

Fig. 8.4 Plot of k_1 against the temperature

SiO_2 as a diluent barely affected the decomposition). The decomposition was found to occur in two stages.

The first stage is characterized by relatively high reaction rates (the heat release is about 2.7 kJ/g of the reagent). In the second stage, decomposition of the condensed residue occurs with significantly lower (by about two orders of magnitude) rates.

Table 8.1 Summarized data on the rate constants for the initial stage of nitrocellulose decomposition

Method	Temperature range, °C	Rate constant, k_1, s^{-1}	Reference
Isothermal calorimetry	140–165	$10^{18}\exp(-47000/RT)$	[1]
Self-inflammation of particles	223–255	$5.6\times10^{18}\exp(-48500/RT)$	[2]
Ignition with gas flux	290–350	$5.5\times10^{18}\exp(-48000/RT)$	[3]
Ignition at contact with hot block	212–242	$2.1\times10^{18}\exp(-48000/RT)$	Sect. 5.3, [4]
Thermography with dilution	182–217	$6\times10^{18}\exp(-47000/RT)$	Sect. 8.1, [5]

The dependence of the effective activation energy on the degree of conversion ($E = 196\,\mathrm{kJ\,mol^{-1}}$ at $\eta = 0.05$–0.2; $E = 125\,\mathrm{kJ\,mol^{-1}}$ at $\eta = 0.6$) indicates that the process is complex. Comparison of literature data on the kinetic constant for the first stage of the decomposition of nitrocellulose, k_1, with experimental data obtained using DTA was of significant interest.

Nitrocellulose decomposition can be described by an equation for a self-catalyzed first-order reaction up to $\eta \approx 0.3$–0.4 [1]. Therefore, the isothermal data (on $\dot{\eta}$ as a function of η up to $\eta = 0.2$) obtained by processing the DTA data were linearized in the $\lg[\dot{\eta}/(1-\eta)] - \eta$ coordinates. The resulting Arrhenius plot for the temperature dependence of k_1 is given in Fig. 8.4.

Kinetic constants for the first stage of the thermal decomposition of nitrocellulose obtained by different methods are presented in Table 8.1. The experimental and literature data are in good agreement.

The temperatures reached in the nitrocellulose self-inflammation and ignition experiments are higher than those that are characteristic of DTA with dilution. However, this method yielded a formal kinetic expression for high-temperature nitrocellulose decomposition ($\dot{\eta} = \dot{\eta}(\eta, T)$), which cannot be obtained using self-inflammation and ignition methods.

8.2 High-Temperature Decomposition of Ammonia Copper Chromate

The copper chromite catalyst SIAC-105, developed at the State Institute of Applied Chemistry (St. Petersburg, Russia), is widely used in the technological process for polyol (C_7–C_{18}) production (in particular, for ester hydrogenation) and as a component in some composite solid propellants. The activity of this catalyst depends on the conditions of the thermal decomposition of the initial reagent (ammonia copper chromate). The kinetic parameters and thermal effect of ammonia copper chromate (ACC) decomposition are important factors in reactor design. They were determined by thermographic experiments using the mechanical dilution of ACC ($MC/mc \approx$

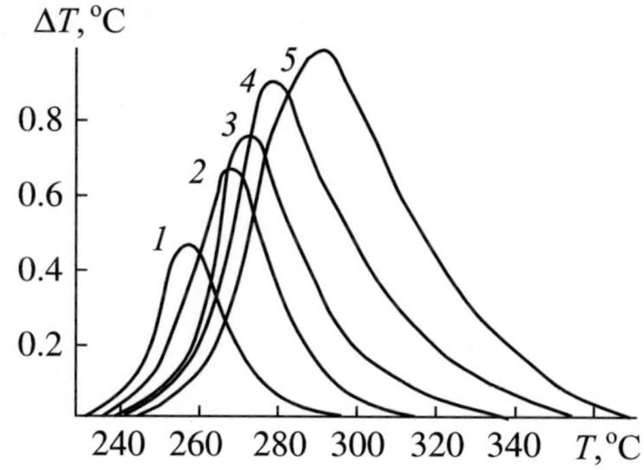

Fig. 8.5 Thermograms of ACC decomposition at $\omega =$ 1.5 (*1*), 2.4 (*2*), 3.4 (*3*), 4.5 (*4*), 6.6 (*5*) deg min^{-1}

100, diluent: burnt Al_2O_3 or SiO_2 powder) in N_2 gas, air and vacuum. DTA-curves for ACC decomposition at a pressure of 13 Pa are presented in Fig. 8.5.

The dependencies $\dot{\eta} = \dot{\eta}(T^{-1})$ corresponding to certain values of η were linearized in semi-logarithmic coordinates (Fig. 8.6). The corresponding effective activation energies were found to be almost constant ($E_{\text{eff}} \approx 112.5\,\text{kJ mol}^{-1}$).

The experimental data were analyzed by a method described in Chap. 6. As a result, it was shown that ACC decomposition for η of up to 50–60% can be quite well described by the following expression for a self-catalyzed first-order reaction:

$$\dot{\eta} = k_1(1-\eta) + k_2\eta(1-\eta).$$

The expressions for the rate constants k_1 and k_2 (Fig. 8.7) are

$$k_1 = 10^{7.7\pm0.8}\exp\left(-\frac{28000 \pm 1500}{RT}\right), \; k_2 = 70k_1. \qquad (8.1)$$

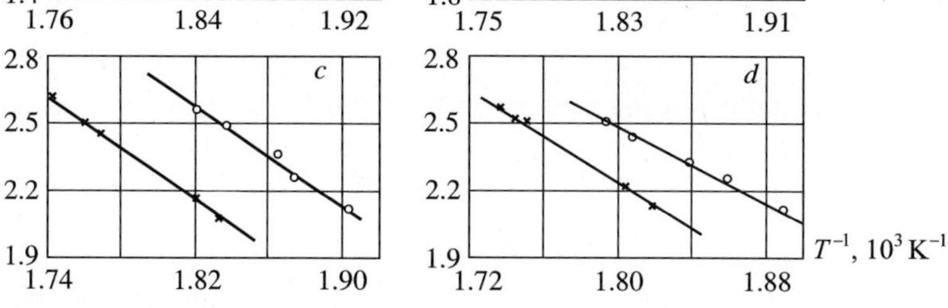

Fig. 8.6 Plots of ACC decomposition rate against the temperature at $\eta = 0.1$ (**a**), 0.3 (**b**), 0.4 (**c**), 0.6 (**d**) in vacuum at $P = 13$ Pa (*circles*) and in N_2 gas at the atmospheric pressure (*crosses*)

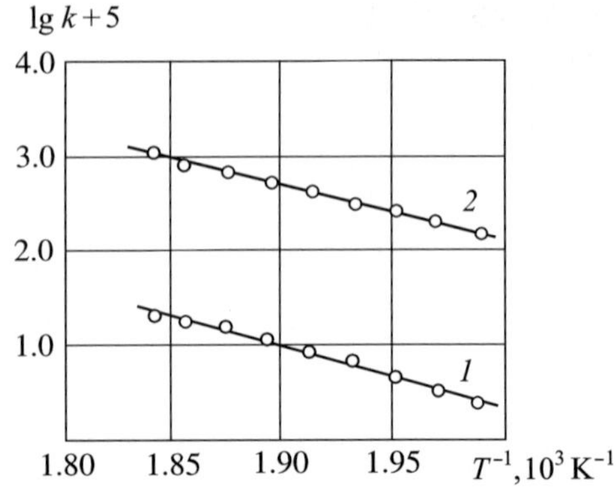

Fig. 8.7 Plots of the decomposition rate constants k_1 *(1)* and k_2 *(2)* against the temperature

In the case of ACC thermal decomposition in N_2 gas and air at atmospheric pressure, the gaseous reaction products inhibit the decomposition, while stimulating gas removal results in accelerated decomposition. For example, the decomposition rate at a pressure of 13 Pa is approximately two times higher than that at atmospheric pressure (Fig. 8.6). This result is in good agreement with data [6], which also indicate that the rate of ACC weight diminution in He flow is significantly higher than those in N_2 gas or Ar. The thermal effects of ACC decomposition in vacuum ($P = 13$ Pa) and N_2 gas are almost identical ($Q = 271 \pm 21\,\mathrm{J\,g^{-1}}$), while in air it is higher due to secondary oxidation reactions ($Q \approx 458\,\mathrm{J\,g^{-1}}$).

8.3 High-Temperature Decomposition of AP

The decomposition of AP has been actively studied for many years. However, most of the data available in the literature corresponds to low-temperature decomposition ($T < 240°$C), while works on high-temperature decomposition are few in number, and the data are often controversial.

8.3.1 Experimental Study of AP High-Temperature Decomposition Kinetics

AP decomposition was studied by the methods of DTA and TGA with thermal dilution in N_2 gas for a sample (10–25 mg) temperature in the range 290–450°C and a heating rate in the range 1–10 deg $\mathrm{min^{-1}}$. The samples were prepared from commercial AP (thermograms obtained for samples prepared from AP doubly recrystallized from water and from commercial AP were identical).

The process rate was found to be significantly affected by the cell material, the pressure in the cell and the conditions of gas removal. The maximum catalytic

Fig. 8.8 Thermographic curves for heat evolution at AP decomposition in working cells made of duralumin (*1*), stainless steel (*2*) and aluminum (*3*) (thermal dilution, $\omega = 6\,\text{deg min}^{-1}$)

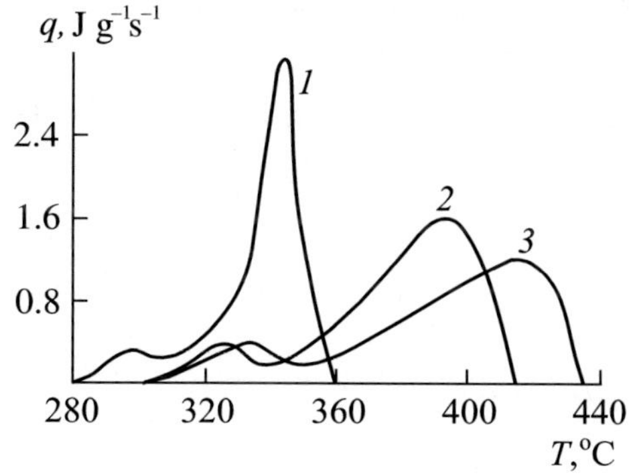

activity (comparing duralumin, stainless steel, Al) was exhibited by duralumin, and the minimum by Al (Fig. 8.8). In the case of the Al cell, the catalytic effect can be neglected since the thermograms obtained are identical to those obtained using the cell in which the sample is isolated from the metal by thin mica plates.

The experimental thermograms and TGA curves are presented in Figs. 8.9 and 8.11. The corresponding dependencies of the nonisothermal reaction rate, $\dot{\eta}$, on the conversion degree, η, are given in Figs. 8.10 (DTA) and 8.12 (TGA).

Data obtained from the DTA and TGA curves for $\eta = \text{const}$ are presented in $\lg\dot{\eta} - (1/T)$ coordinates (Fig. 8.13). Jacobs and Russel-Jones found that the AP particle size does not influence the decomposition rate at temperatures above the phase transformation point (240°C) [7]. The temperature dependencies of $\dot{\eta}$ at $\eta =$ const for uniform AP samples with effective particle sizes of 200 μm and 50 μm are given in Fig. 8.13. One can see that the AP particle size (at least, within the indicated limits) does not affect the AP high-temperature decomposition rate, which agrees with the literature data mentioned above. The values of the effective decomposition activation energy obtained by analyzing the data shown in Figs. 8.10 and 8.12 are almost constant and equal to $127\text{–}133\,\text{kJ mol}^{-1}$ for the entire range of the degree of conversion. The constant nature of the effective activation energy indicates that the analytical approach used is correct (see Chap. 6 and review by Merzhanov [8]).

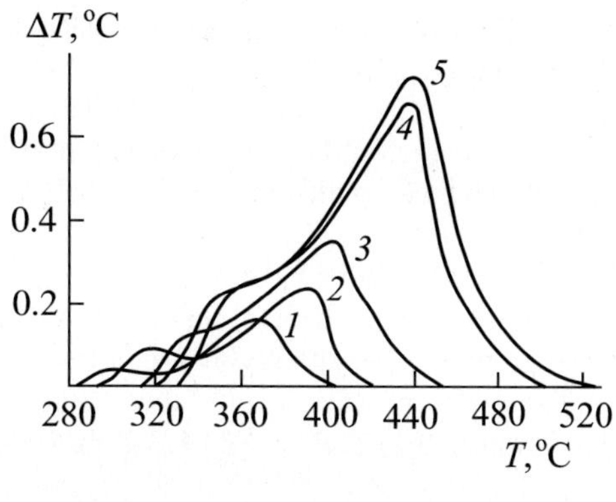

Fig. 8.9 Thermograms of the high-temperature decomposition of AP at $\omega = 1.05$ (*1*), 1.5 (*2*), 5.0 (*3*), 7.4 (*4*), 10.0 (*5*) deg min^{-1}

Fig. 8.10 Plots of the nonisothermal rate of AP decomposition (DTA with thermal dilution) against the conversion degree at $\omega = 10.0$ (*1*), 7.4 (*2*), 5.0 (*3*), 1.5 (*4*), 1.05 (*5*) deg min^{-1}

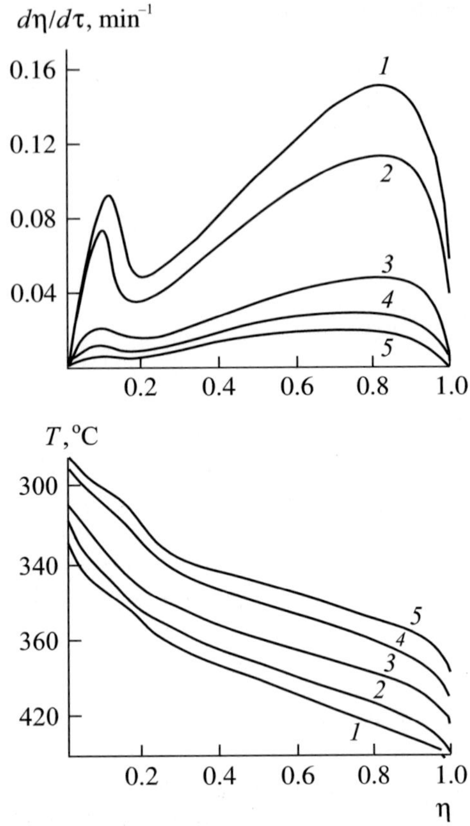

The nonisothermal reaction rates obtained by DTA and TGA are in a very good agreement within the experimental accuracy (Fig. 8.13).

The results of the analysis of the experimental nonisothermal data and the $\dot{\eta}(\eta)$ dependencies (Fig. 8.14) recalculated for isothermal conditions (see Sect. 4.3) clearly indicate the existence of two stages in AP high-temperature decomposition. The first stage ("fast") is characterized by relatively low conversion degrees ($\eta < 0.1$–0.15), while the second one ("slow") takes place at higher conversion degrees ($\eta > 0.15$–0.18). The "fast" decomposition can be described by a kinetic equation for a first-order reaction with activation energy $E_1 = 133 \pm 4\,\mathrm{kJ\,mol^{-1}}$. The data on k_1 (Fig. 8.15) obtained are in a satisfactory agreement with the literature data [10, 9] obtained by different (isothermal) methods based on weight decrease and the evolution of gas.

The expression for k_1 remains constant at temperatures from 200 to 450°C, which implies that the mechanism for the "fast" AP decomposition does not change in this temperature range, although the contribution from each stage changes.

The reaction order of the "slow" stage is 0.5 (Table 8.2 and Fig. 8.16), and its activation energy is $E_2 = 133 \pm 4\,\mathrm{kJ\,mol^{-1}}$ (Fig. 8.17).

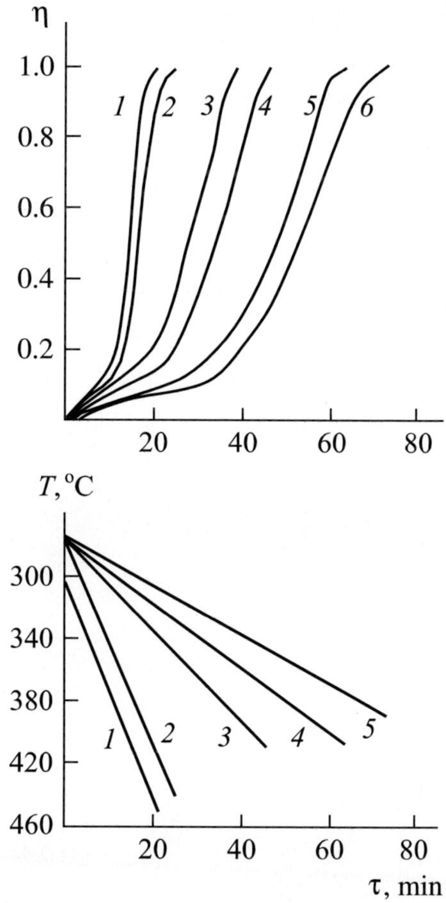

Fig. 8.11 Thermogravimetric curves for the decomposition of AP at $\omega = 6.3$ (*1*), 5.8 (*2*), 3.6 (*3*), 2.9 (*4*), 2.1 (*5*), 1.5 (*6*) deg min^{-1}

The data on k_2 obtained by the DTA and TGA methods were compared with literature data obtained by methods based on linear pyrolysis (Chap. 3) and the weight decrease under isothermal conditions [9]. Jacobs and Russel-Jones used a so-called "contracting sphere" equation

$$1-(1-\eta)^{0.5}=k_2t \tag{8.2}$$

to describe the second stage thermal decomposition kinetics (at 305–380°C) on the basis of data on the decrease in the sample weight [11]. It can be shown that the only difference between Eq. (8.2) and the expression

$$\dot{\eta}=k_2(1-\eta)^{0.5} \tag{8.3}$$

used in our work is the constant factor 0.5 on the right hand side of (8.2). Taking this factor into account, one can see that the data on k_2 given in [11] are in good agreement with those in our work (Fig. 8.17, plots 6 and 7).

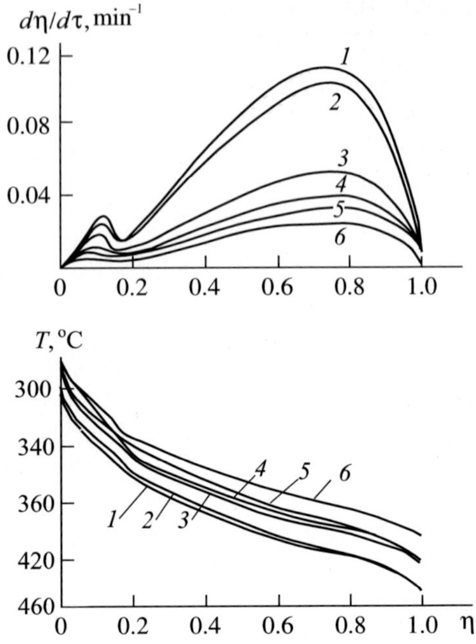

Fig. 8.12 Plots of the non-isothermal rate of AP decomposition (TGA with thermal dilution) against the conversion degree at $\omega = 6.3$ (*1*), 5.8 (*2*), 3.6 (*3*), 2.9 (*4*), 2.1 (*5*), 1.5 (*6*) deg min^{-1}

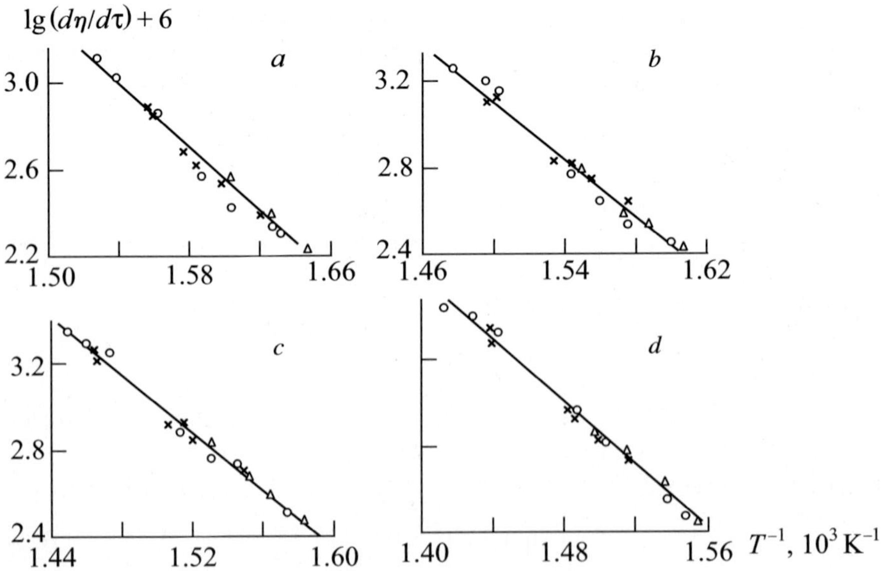

Fig. 8.13 Plots of the AP decomposition rate against the temperature at $\eta = 0.1$ (**a**), 0.3 (**b**), 0.5 (**c**), 0.7 (**d**); DTA, $d_{AP} = 200\,\mu m$ (*circles*); TGA, $d_{AP} = 200\,\mu m$ (*crosses*); TGA, $d_{AP} = 50\,\mu m$ (*triangles*)

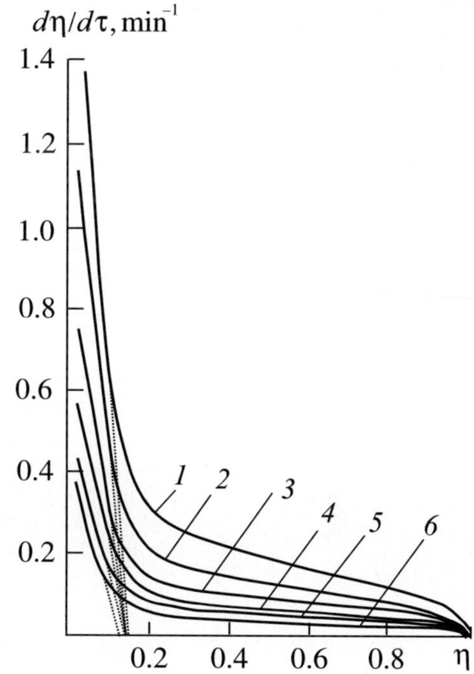

Fig. 8.14 Plots of the isothermal rates of high-temperature decomposition of AP at $T = 420$ (*1*), 410 (*2*), 400 (*3*), 390 (*4*), 380 (*5*), 370 (*6*) °C

Data on the rate constants for "complete" conversion, calculated from the experimental data obtained by linear pyrolysis (see Fig. 3.4), are also plotted in Fig. 8.17. The linear pyrolysis rate, U, is related to the rate constant of the limiting stage of the volumetric decomposition, k_{lp}, as

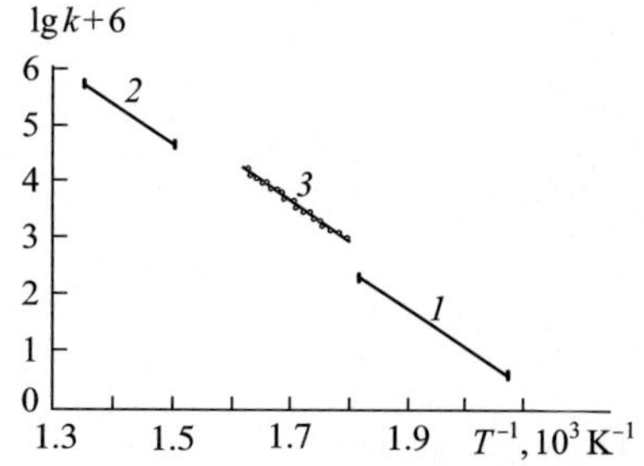

Fig. 8.15 Plots of the rate constant, k_1, of the first ("fast") stage of the high-temperature decomposition of AP against the temperature based on data obtained in [9] (*1*); [10] (*2*); and data in this work (*3*)

Table 8.2 Rate constants, k_2, for the second ("slow") stage of the high-temperature decomposition of AP

Temperature, °C	370	380	390	400	410	420
K_2 (s^{-1})$\times 10^3$	0.80	1.18	1.66	2.40	3.45	4.78
Reaction order	0.49	0.51	0.53	0.52	0.51	0.51

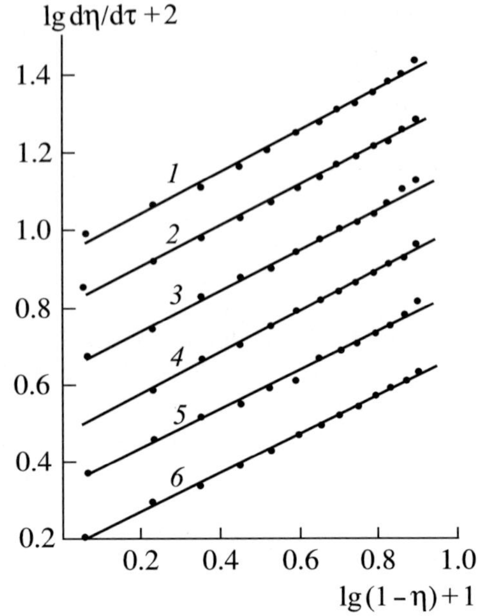

Fig. 8.16 Linearization of the isothermal curves for the second ("slow") stage of the high-temperature decomposition of AP at $T = 420$ (*1*), 410 (*2*), 400 (*3*), 390 (*4*), 380 (*5*), 370 (*6*) °C

$$U = k_{\mathrm{lp}}(T_S, \eta_S)\delta \frac{\mathrm{R}T_S^2}{E(T_S - T_\infty)}, \tag{8.4}$$

where δ is the preheat layer thickness and η_S is the degree of conversion on the pyrolysis surface.

The value of η_S cannot be determined from experimental data obtained by linear pyrolysis (that is, from the $U(T_\mathrm{S})$ dependence) within the existing linear pyrolysis theory. However, one can estimate η_S with good accuracy from experimental data

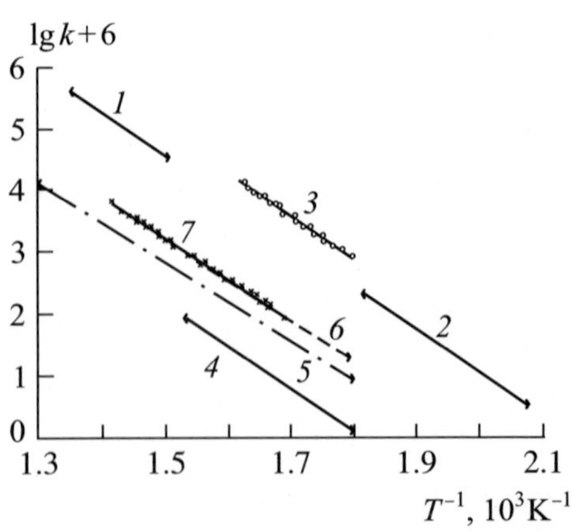

Fig. 8.17 Summarized data on the kinetics of the high-temperature decomposition of AP for the "fast" stage (data obtained in [9] (*1*); [10] (*2*); and data in this work (*3*)) and the "slow" stage (data obtained in [9] (*4*); data obtained by the linear pyrolysis method without recalculation for $n = 0.5$, where the finite character of η_S during the linear pyrolysis of AP is not taken into account (*5*); data obtained in [11] (*6*); data on DTA-TGA shown in Figs. 8.13 and 8.14 (*7*))

on linear pyrolysis using available data on the reaction order ($n = 0.5$) and k_2 for the "slow" stage of high-temperature AP decomposition obtained by DTA/TGA. The corresponding expression is

$$\eta_S = 1 - \frac{1}{1-\alpha_1}\left(\frac{k_{\mathrm{lp}}}{k_2}\right)^2, \tag{8.5}$$

where $\alpha \approx 0.1$ is the fraction of the reagent reacting during the first ("fast") stage (see Sect. 8.3.2).

In the temperature range 360–440°C, $\frac{k_{\mathrm{pl}}}{k_2} \approx 0.3 - 0.4$ and the degree of conversion on the surface for AP linear pyrolysis is $\eta_S \approx 0.82$–0.90. The unreacted fraction of AP leaves the linear pyrolysis surface in the form of ammonia and perchloric acid (the products of dissociative sublimation at thermodynamic equilibrium that occurs in parallel with the irreversible decomposition of AP).

8.3.2 Reaction Scheme and Equation for High-Temperature Decomposition of AP

Let us consider the high-temperature decomposition of AP that occurs via two independent parallel reactions. In a first approximation, the fraction of the reagent decomposing due to the first ("fast") stage is constant ($\alpha_1 = \mathrm{const}$), and one can write [12]:

$$\alpha_1 A \xrightarrow{k_{01};E_1;n_1;Q_1} B, \tag{8.6}$$

$$(1-\alpha_1)A \xrightarrow{k_{02};E_2;n_2;Q_2} B, \tag{8.7}$$

$$\begin{aligned} -\frac{dC_1}{dt} &= k_1(T)C_1^{n_1} \\ -\frac{dC_2}{dt} &= k_2(T)C_2^{n_2} \end{aligned}, \tag{8.8}$$

where

$$C_{1,2} = \frac{m_{1,2}}{m_0} = 1 - \frac{Q_{1,2}}{Q},$$

$t = 0$; $C_1 = \alpha_1$; $C_2 = 1 - \alpha_2$; $E_1 = E_2 = (133 \pm 4)\,\mathrm{kJ\,mol^{-1}}$, $k_{01} = 2.2 \times 10^9\,\mathrm{s^{-1}}$; $n_1 = 1$; $k_{02} = 4.4 \times 10^7\,\mathrm{s^{-1}}$; $n_2 = 0.5$.

For the total reaction rate recorded in the experiment, one can write

$$-\frac{dC}{dt} = k_1(T)C_1 + k_2(T)\sqrt{C_2}. \tag{8.9}$$

From Eq. (8.8)

$$\frac{dC_2}{dC_1} = \eta_0 \frac{\sqrt{C_2}}{C_1}, \text{ where } \eta_0 = \frac{k_2(T)}{k_1(T)},$$

and

$$\frac{dC_2}{\sqrt{C_2}} = \eta_0 \frac{dC_1}{C_1}. \tag{8.10}$$

After integration one obtains

$$2(\sqrt{C_2} - \sqrt{1-\alpha_1}) = \eta_0 \ln \frac{C_1}{\alpha_1}. \tag{8.11}$$

Let $C_2 = Z$ and $\eta = 1 - (C_1 + C_2)$, then $C_1 = 1 - \eta - Z$ and

$$\sqrt{Z} - \sqrt{1-\alpha_1} = \frac{\eta_0}{2} \ln(1-\eta-Z) - \frac{\eta_0}{2} \ln \alpha_1. \tag{8.12}$$

Since $-\frac{dC}{dt} = \frac{d\eta}{dt}$,

$$\frac{d\eta}{dt} = k_1(T)(1-\eta-Z+\eta_0\sqrt{Z}). \tag{8.13}$$

After substituting $k_1(T)$, α and η_0 into Eq. (8.9), one obtains the desired kinetic equation for the high-temperature decomposition of AP:

$$\frac{d\eta}{dt} = 2.2 \times 10^9 \exp\left(-\frac{32000}{RT}\right)(1-\eta-Z+0.02\sqrt{Z}), \tag{8.14}$$

where Z is the root of the transcendental equation

$$\sqrt{Z} = 0.01 \ln(1-\eta-Z) + 0.93.$$

8.3.3 Effect of Dissociative Sublimation and Gas Removal Conditions on the Macrokinetics of High-Temperature Decomposition of AP

The decomposition of AP is characterized by two parallel processes: an irreversible reaction (I) and dissociative sublimation at thermodynamic equilibrium (II). At moderate temperatures, the influence of process (II) on the decomposition of AP can be neglected due to the low vapor pressures of perchloric acid and ammonia, the products of process (II). At the high temperatures ($T \approx 350$–$500°C$) characteristic of DTA and TGA with dilution [12], as well as for linear pyrolysis and burning of AP, the role of sublimation becomes more significant.

Let us analyze the influence of dissociative sublimation on the macrokinetics of the high-temperature decomposition of AP and model AP-based mixtures. In the analysis, ideas about the effect of the "evaporation in volume" that takes place during the decomposition of a condensed phase of volatile explosives are used [3]. In spite of the fact that at high temperatures (up to 550°C) AP does not melt, its decomposition is generally similar to the decomposition of secondary explosive melts. In the decomposition of AP, the unreacted fraction of the reagent undergoes sublimation, while in the decomposition of explosives it evaporates. Beyond the point of lattice transformation from orthorhombic to cubic (at 240°C), the porosity of AP significantly increases. The sublimation products located in the microcracks formed are in equilibrium with the solid phase.

In the experiments with thermal dilution (when a thin sample layer is pressed between two heat-conducting metal blocks), gas removal is hindered. Under these conditions, the gaseous products of process (I) are saturated with the vapors of perchloric acid and ammonia, the products of process (II). For the sake of brevity, the products of irreversible decomposition will be called the "gas," while the products of dissociative sublimation will be termed the "vapor" here. The process rate, $\dot{q}$ (recorded in calorimetry and thermography), is equal to the difference between the rate of heat evolution due to process (I) and the rate of heat absorption due to process (II).

In thermogravimetric experiments, the rate of the decrease in the sample weight, $\dot{m}$, is the sum of the rates of gas ($\dot{m}_g$) and vapor ($\dot{m}_v$) formation:

$$\dot{q} = Q\dot{m}_g \left(1 - \frac{\Delta H}{Q}\frac{\dot{m}_v}{\dot{m}_g}\right), \tag{8.15}$$

$$\dot{m} = \dot{m}_g \left(1 + \frac{\dot{m}_v}{\dot{m}_g}\right), \tag{8.16}$$

where Q is the thermal effect of the reaction and ΔH is the heat of sublimation. At a constant temperature T, one can write for $\dot{m}_v/\dot{m}_g$:

$$\frac{\dot{m}_v}{\dot{m}_g} = \frac{\mu_v}{\mu_g}\frac{P_{sv}(T)}{P_\infty - P_{sv}(T)}, \tag{8.17}$$

where μ_v and μ_g are the molecular weights of the vapor and the gas, respectively, $P_{sv}(T)$ is the saturated vapor pressure and P_∞, is the ambient pressure.

For linear heating of the sample,

$$\frac{\dot{m}_v}{\dot{m}_g} = \frac{\mu_v}{\mu_g}\frac{P_{sv}(T)}{P_\infty - P_{sv}(T)}\left[1 + \frac{m_g}{\dot{m}_g}\frac{\Delta H\omega}{\mathrm{R}T^2}\frac{P_{sv}(T)}{P_\infty - P_{sv}(T)}\right], \tag{8.18}$$

where m is the mass of gas in the sample pores under quasi-stationary conditions and ω is the heating rate.

At $\omega \le 10\,\text{deg}\,\text{min}^{-1}$ (characteristic of standard thermographic experiments on AP decomposition), the second term in the square brackets in Eq. (8.18) is much less than unity. The expression for the relationship between the heat evolution rate, $\dot{q}$, and the decomposition rate, $\dot{\eta}$, is identical to that for the isothermal process:

$$\dot{q} = \dot{\eta} Q \left[1 - \frac{\Delta H}{Q} \frac{\mu_v}{\mu_g} \frac{P_{sv}(T)}{P_\infty - P_{sv}(T)} \right]. \tag{8.19}$$

Equation (8.19) can be integrated over the interval T_{in}–T_{fin}, where T_{in} and T_{fin} are the initial and final temperatures of the thermographic experiment. The conditions required to make the integration valid are

$$\frac{dT}{dt} = \omega = \text{const} \tag{8.20}$$

and

$$P_{sv}(T) << P_\infty \tag{8.21}$$

over the entire temperature range ($T_{in} - T_{fin}$).

Then one can write a relationship between the experimentally determined and the true decomposition thermal effects (Q_{exp} and Q) as

$$Q_{exp} = Q - \frac{\Delta H \mu_v}{P_\infty \omega \mu_g} \int_{T_{in}}^{T_{fin}} \dot{\eta}(T) P_{sv}(T) dT. \tag{8.22}$$

The integral on the right hand side of Eq. (8.22) weakly depends on P_∞, and so at $\omega = \text{const}$

$$Q_{exp} = Q - \frac{\text{const}}{P_\infty}. \tag{8.23}$$

In thermogravimetric experiments on the decomposition of thermally diluted AP samples at various pressures of N_2 gas, the dependence $Q_{exp}(P_\infty)$ was recorded. The data obtained in two sets of experiments that employed working cells of different designs are presented in Table 8.3. The necessary condition (8.20) was met with the required accuracy due to thermal dilution. The obtained thermograms were compared with the temperature dependence of the vapor pressure of the products of the dissociative sublimation of AP [14]. As a result, the condition (8.21) was found to be rigorously met at $\omega = 3$–$4\,\text{deg min}^{-1}$ and $P_\infty \geq 3 \times 10^4$ Pa.

The data presented in Table 8.3 were analyzed using Eq. (8.23) in $Q_{exp} - P_\infty^{-1}$ coordinates (Fig. 8.18). The value obtained for the true thermal effect of the high-temperature decomposition of AP ($Q = 1.16\,\text{kJ g}^{-1}$) is in good agreement with the value obtained in [15] using a differential scanning calorimeter and vials equipped with ball clacks for pressures of up to 1 MPa ($Q = 1.14\,\text{kJ g}^{-1}$).

From Eqs. (8.16) and (8.17), one can see that as the temperature increases, the effective energy of the process resulting in the disappearance of the AP condensed

Table 8.3 Dependence of the thermal effect of the high-temperature decomposition of AP, Q_{exp}, on the pressure, P_∞

$P_\infty \times 10^5$ (Pa)	0.01	0.3	0.5	1.0	10.0
Q_{exp}, kJ g^{-1}, cell #1	–	0.91	1.00	1.06	1.17
Q_{exp}, kJ g^{-1}, cell #2	0.74	0.83	0.94	1.03	1.15

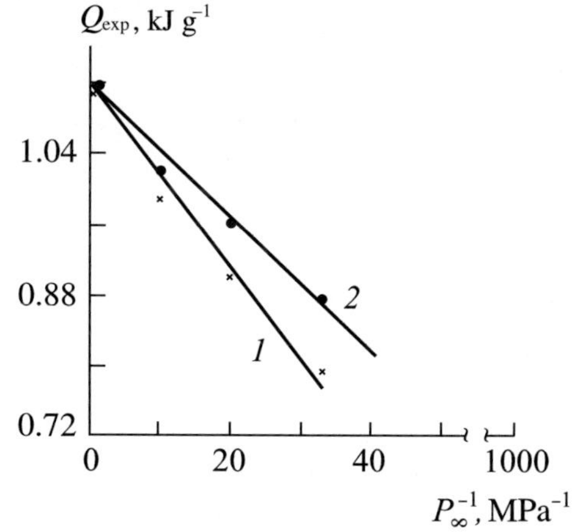

Fig. 8.18 Determination of the thermal effect of the high-temperature decomposition of AP in working cells # 1 and #2

phase remains equal to the true activation energy of the irreversible decomposition until

$$\frac{\mu_v}{\mu_g}\frac{P_{sv}(T)}{P_\infty - P_{sv}(T)} << 1 \tag{8.24}$$

is valid.

At higher temperatures, the fraction of the sublimation products increases and, according to Eq. (8.16), the effective activation energy increases, approaching infinity for $P_{sv}(T) = P_\infty$. This was observed in experiments on the linear pyrolysis of AP accompanied by the burning of gaseous pyrolysis products (Fig. 3.5).

References

1. Manelis GB, Rubtsov YuI, Smirnov LP, Dubovitsky FI (1962) Kinet Catal 1:32
2. Grigor'ev YuM, Maksimov EI, Merzhanov AG (1965) Combust Explos Shock Waves 1:70
3. Lisitsky VI (1970) Combust Explos Shock Waves 6:494
4. Shteinberg AS, Ulybin VB, Barzykin VV, Merzhanov AG (1966) J Eng Phys Thermophys 4:282
5. Goncharov EP, Driatskaya GI, Merzhanov AG, Shteinberg AS (1971) Dokl Chem 197:254
6. Pashkova TL, Tyuryaev IYa, Goncharov EP, Shteinberg AS (1973) In: Tyuryaev IYa (ed) Catalysts of basic organic syntheses. Khimiya, St. Petersburg, p 34
7. Jacobs PW, Russel-Jones A (1969) J Phys Chem 72:202
8. Merzhanov AG (1973) Combust Explos Shock Waves 9:4
9. Manelis GB, Rubtsov YuI (1966) Rus J Phys Chem 40:416
10. Agrawal JP (1998) Prog Energy Combust Sci 24:1
11. Jacobs PW, Russel-Jones A (1967) AIAAJ 5:829
12. Goncharov EP, Merzhanov AG, Shteinberg AS (1972) In: Stesik LN (ed) Combustion and explosion. Nauka, Moscow, p 765
13. Merzhanov AG, Slutsker BM, Shteinberg AS (1968) Combust Explos Shock Waves 4:312
14. Jnami S, Rosser W, Weis NM (1963) J Phys Chem 67:1077
15. Santacesaria S, Giuffre L, Fusi G (1971) Rev Combust 25:376

Chapter 9
Kinetic of Deep High-Temperature Decomposition of Model and Commercial AP-Based Composite Solid Propellants

Abstract The kinetics of fast high-temperature decomposition is quantitatively studied for a number of model and commercial heterogeneous polymer–ammonium perchlorate systems using original nonisothermal kinetics methods employing DTA, TGA and DSC instruments. Maximum temperatures corresponding to the determined kinetic parameters reach 400–450°C, which are quite close to those in the solid propellant reaction zone near the burning surface. Data presented in this chapter can be used for the analysis of combustion patterns and mechanisms of solid propellants, while the results of earlier works on the kinetics of low-temperature decomposition are only useful for assessing the stabilities of the propellants during long-term storage. This is also confirmed by comparing the corresponding kinetic constants. For example, the activation energies for the low-temperature decomposition of the majority of the studied solid propellants are close to 120 kJ mol^{-1}, which corresponds to the decomposition of AP. The activation energy of the second stage of the high-temperature decomposition of almost all AP–polymer mixtures is twice as high. An explanation of this quite general effect is presented. An excellent correlation between the obtained kinetic constants of high-temperature reactions and the ignition parameters and the rates of subatmospheric burning for the studied solid propellants is discussed.

9.1 Kinetics of High-Temperature Decomposition of AP–Polymer Model Systems

The macrokinetics of the high-temperature decomposition of AP-based heterogeneous model and commercial mixtures were studied by DTA and TGA methods using thermal dilution. In most of the experiments, the oxidizer particle size was $d_{AP} = 200\,\mu m$, while in other experiments $d_{AP} = 50\,\mu m$.

Since the fuel-binder content in commercial composite solid propellants is about 20%, the oxidizer/fuel ratio in the model mixtures was 4/1. The following polymer powders were used as fuels in the model mixtures: polystyrene (PS), polyvinylchloride (PVC), polyacrylonitrile (PAN) and polyethylene (PE). The particle size of the powders was not important, since all of these materials melt before active decomposition of their mixtures with AP begins. Some characteristics of the commercial NG-2-60-DTM-grade composite solid propellants (cast materials based on polyurethane binders synthesized from polyoxipropylenepolyols) studied by ignition methods are given in Chap. 5.

Since the decomposition kinetics for the first four polymers has been thoroughly analyzed by Grassi [1] and Madorsky [2], and for polyurethanes by Benin and coworkers [3], we focused mostly on the experimental study of the high-temperature decomposition of AP-based mixtures. For the sake of brevity, the original thermograms are omitted (except for the ones for the AP–PS and HG-3-60-DTM systems); they look quite similar anyway.

9.1.1 AP–PS System

The original thermograms obtained under thermal dilution using installation T-2 are shown in Fig. 9.1.

From the thermograms and the results of analyzing them (Figs. 9.2–9.4), one can conclude that the decomposition of the AP–PS system occurs via two stages with total thermal effect $Q = 2.5 \pm 0.2\,\mathrm{kJ\,g^{-1}}$. This conclusion is based on the monotonous change in the effective activation energy as the conversion degree in increased (Fig. 9.4). The rate constants and effective activation energy ($E_1 = 133\,\mathrm{kJ\,mol^{-1}}$) of the first stage of the AP–PS decomposition (at $\eta < 0.15$) are close to those of AP

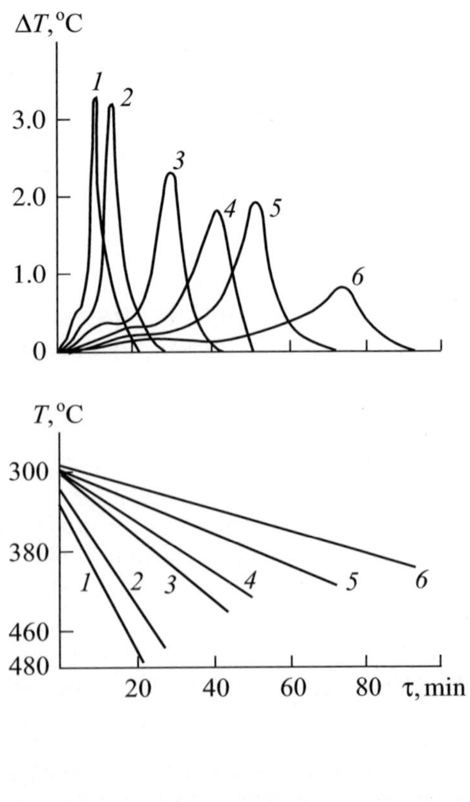

Fig. 9.1 Thermograms of AP–PS mixture decomposition at heating rate $\omega = 7.6$ (*1*), 6.1 (*2*), 3.0 (*3*), 2.6 (*4*), 1.6 (*5*) and 1.1 (*6*) °C min^{-1}

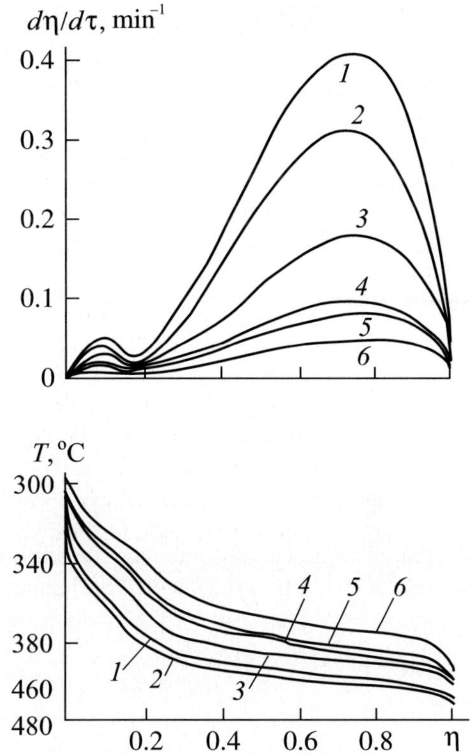

Fig. 9.2 Plots of the nonisothermal rate of AP–PS mixture decomposition against the conversion degree at heating rate $\omega = 7.6$ (*1*), 6.1 (*2*), 3.0 (*3*), 2.6 (*4*), 1.6 (*5*) and 1.1 (*6*) °C min^{-1}

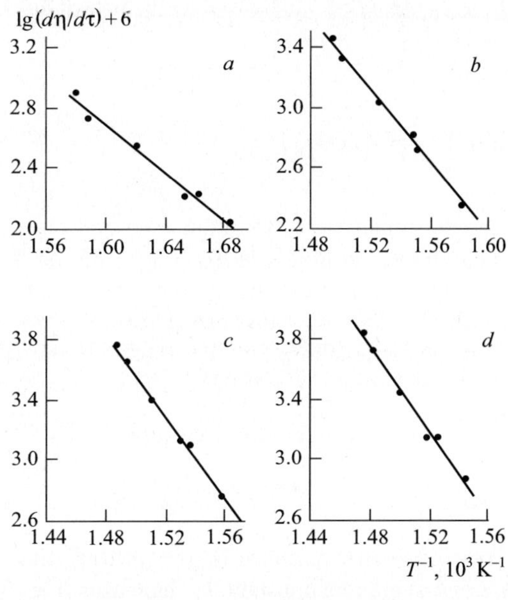

Fig. 9.3 Temperature dependence of the rate of AP–PS mixture decomposition at conversion degree $\eta = 0.1$ (**a**), 0.3 (**b**), 0.5 (**c**) and 0.7 (**d**)

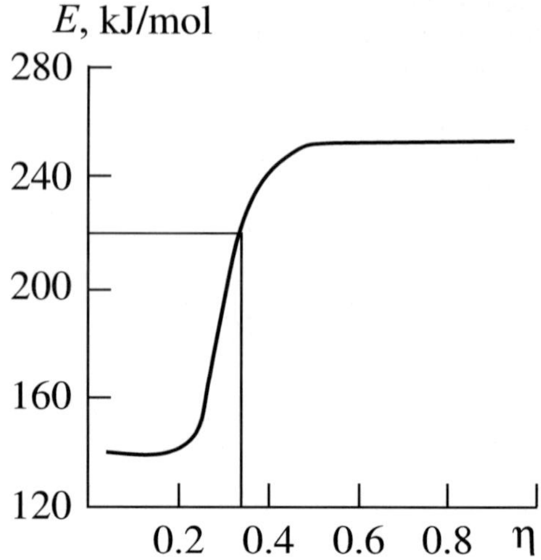

Fig. 9.4 Plot of the effective activation energy of AP–PS mixture decomposition against the conversion degree

decomposition. The second-stage reaction rate depends weakly on the conversion degree (at $0.4 < \eta < 0.8$), which is characteristic of pseudo-zero order reactions. In this case, the expression for the rate constant is

$$k_2 = 10^{18.8\pm0.7} \exp\left(-\frac{63000 \pm 2300}{RT}\right) \left[s^{-1}\right]. \tag{9.1}$$

It is clear that the kinetic parameters and kinetic equation for the second stage of the decomposition of the AP–PS system differ significantly from those for the second stage of the high-temperature decomposition of "pure" AP. This situation was found to be quite characteristic of the second-stage decomposition of AP–polymer mixtures.

9.1.2 AP–PE System

The dependencies of the nonisothermal rates of the decomposition of the AP–PE mixture (AP particle size $d_{AP} = 200\,\mu m$) on the conversion degree are presented in Fig. 9.5. In this case, the thermal effect of the process was found to be $Q = 2.3 \pm 0.2\,kJ\,g^{-1}$. The most interesting feature of the system is the single-stage character of its decomposition. For $\eta = 0.05$–0.9, the effective activation energy is constant ($E = 250\,kJ\,mol^{-1}$), and the process can be satisfactorily described by an equation for a first-order reaction with reaction constant

$$k = 10^{16.9\pm1.0} \exp\left(-\frac{60000 \pm 2600}{RT}\right) \left[s^{-1}\right]. \tag{9.2}$$

For smaller AP particles ($d_{AP} = 50\,\mu m$), the process becomes two-stage, and the second-stage rate constant, k_2, increases (Fig. 9.6).

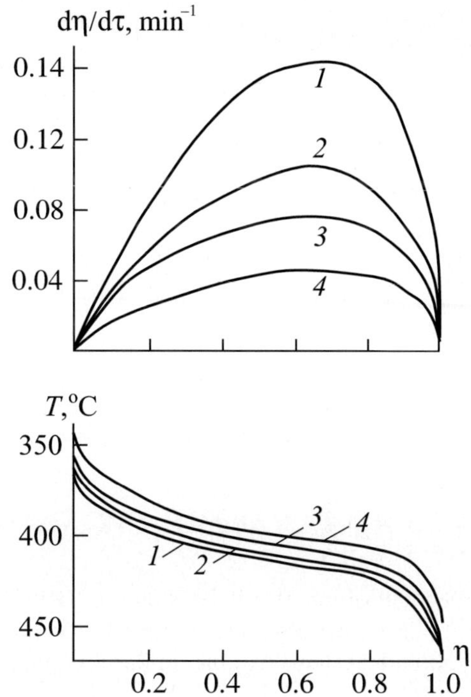

Fig. 9.5 Plots of the nonisothermal rate of AP–PE mixture decomposition ($d_{AP} = 200\,\mu m$) against the conversion degree at heating rate $\omega = 5.7$ (*1*), 4.3 (*2*), 2.1(*3*), 1.7 (*4*) °C min^{-1}

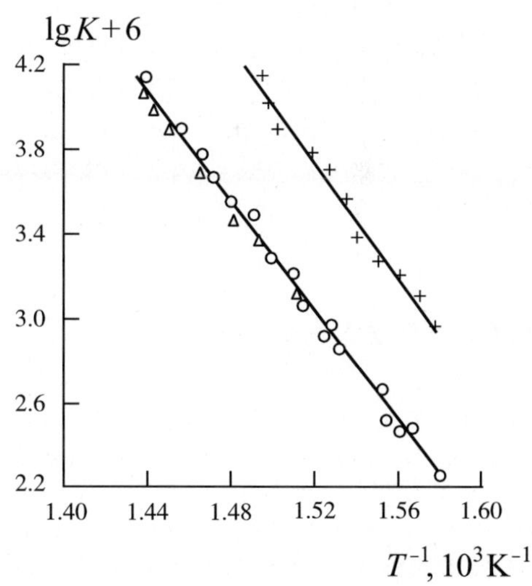

Fig. 9.6 Plots of the rate constants for AP–PE mixture decomposition against the temperature for samples characterized by $d_{AP} = 50\,\mu m$ (*crosses*) and $200\,\mu m$ (*circles and triangles*)

Data on the kinetic constants for the high-temperature decomposition of AP–PE mixtures that differ in terms of AP particle size are given in Table 9.1.

Based on these data (taking into account that the AP powders used were not strictly monodisperse), one concludes that the second-stage rate constant is proportional to the specific area of the polymer–oxidizer contact:

Table 9.1 The effect of AP particle size on the kinetics of the high-temperature decomposition of the AP–PE mixture

Temperature, °C	Rate constant, s^{-1}		$\frac{k_2(d_{AP}=50\,\mu m)}{k(d_{AP}=200\,\mu m)}$
	k ($d_{AP}=200\,\mu m$)	k_2 ($d_{AP}=50\,\mu m$)	
340	2.66×10^{-5}	1.09×10^{-4}	4.26
360	1.27×10^{-4}	5.5×10^{-4}	4.33
380	5.5×10^{-4}	2.5×10^{-3}	4.55
400	2.18×10^{-3}	1.04×10^{-2}	4.77
420	8×10^{-3}	3.98×10^{-2}	4.98

$$k_2 \sim S_{spec} \sim \frac{1}{d_{AP}}. \tag{9.3}$$

9.1.3 AP–PAN and AP–PVC Systems

Decomposition of AP–PAN and AP–PVC model systems ($d_{AP}=200\,\mu m$) is quite similar to that observed for an AP–PS mixture: the reactions occur via two stages, and both systems exhibit similar total thermal effects ($Q_{AP\text{-}PAN}=2.3\pm0.2\,kJ\,g^{-1}$ and $Q_{AP\text{-}PVC}=2.2\pm0.2\,kJ\,g^{-1}$). Approximately 10% of the total heat is released during the first stage.

For both mixtures, the second-stage process (Fig. 9.7) can be described by an equation for a first-order reaction. The rate constants are:

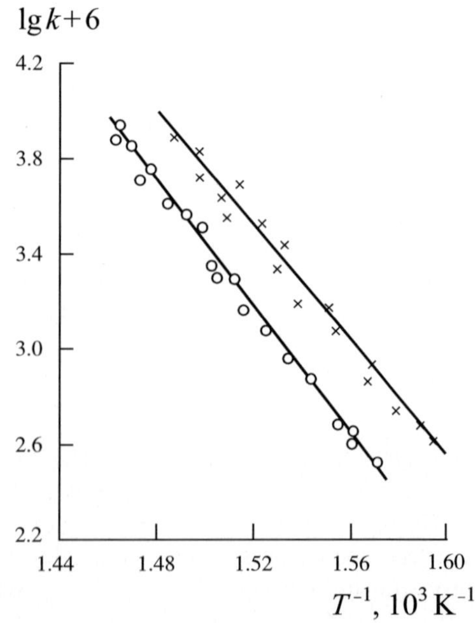

Fig. 9.7 Plots of the decomposition rate constants for AP–PAN (*crosses*) and AP–PVC (*circles*) mixtures against the temperature

$$k_2 = 10^{18.6\pm1.1} \exp\left(-\frac{60100 \pm 3300}{RT}\right) \left[s^{-1}\right] \text{ for the AP–PAN system,} \quad (9.4)$$

$$k_2 = 10^{18.2\pm0.2} \exp\left(-\frac{62000 \pm 2900}{RT}\right) \left[s^{-1}\right] \text{ for the AP–PVC system.} \quad (9.5)$$

9.2 Kinetics of High-Temperature Decomposition of Polyurethane-Based Composite Solid Propellants

The high-temperature decomposition of NG-2-60-DTM composite solid propellants (samples 1 and 2) developed at the State Institute of Applied Chemistry (St. Petersburg, Russia) was studied by the method of ignition using a hot block (Sect. 5.4) and by DTA and TGA with thermal dilution. The Al-free samples 1 and 2 consisted of 80 wt% AP and 20 wt% fuel-binder (a toluylenediisocyanate-cured mixture of polyoxipropylenepolyols). Sample 2 contained a combustion catalyst (0.5% Fe_2O_3).

Original thermograms for sample 1 are given in Fig. 9.8 and the data obtained by analyzing them are shown in Figs. 9.9 and 9.10. Thermograms for sample 2 are qualitatively similar to those for sample 1 (the working temperatures corresponding to sample 1 decomposition are 30–40°C lower than those for sample 2 decomposition, see Fig. 9.11). Both samples were found to decompose via two stages. A comparison of the DTA and TGA data for each stage can give some interesting results. For

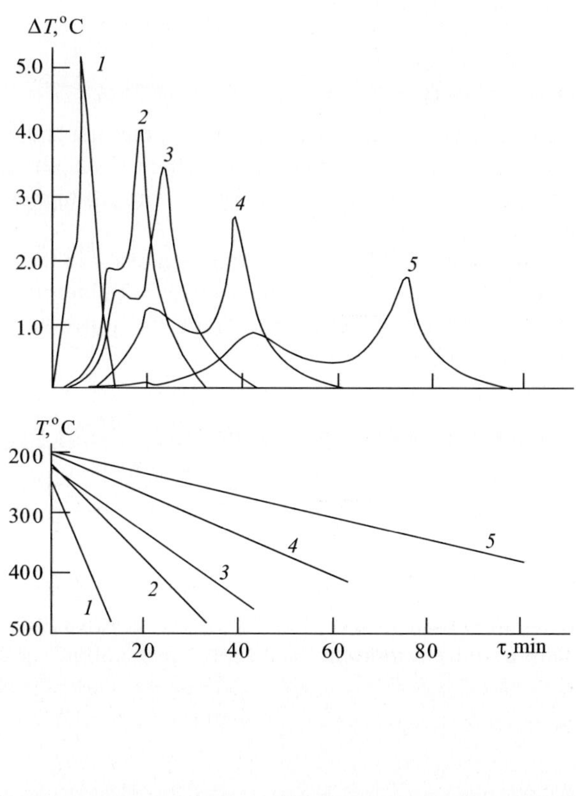

Fig. 9.8 Thermograms of NG-2-60-DTM (sample 1) decomposition at heating rate $\omega = 14.1$ (*1*), 8.8 (*2*), 5.6 (*3*), 3.4 (*4*), 2.6 (*5*) °C min^{-1}

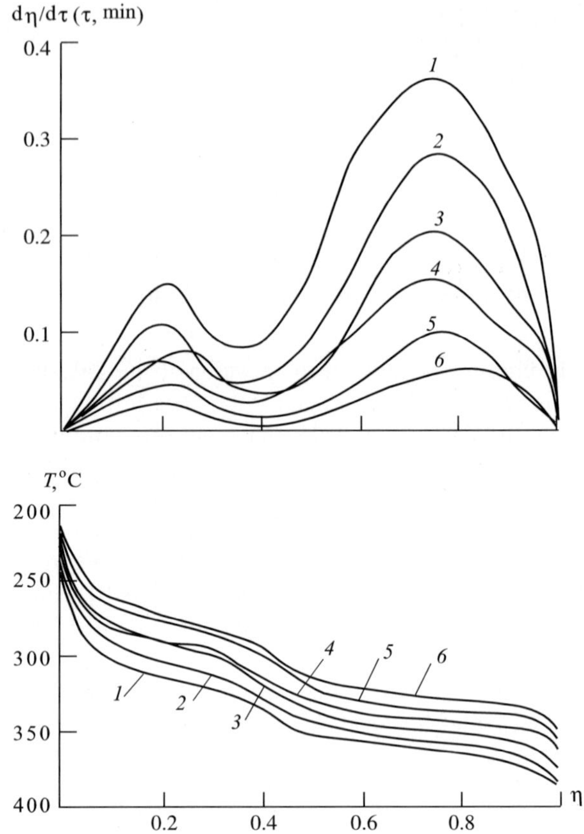

Fig. 9.9 Plots of the nonisothermal rate of NG-2-60-DTM (sample 1) decomposition against the conversion degree at heating rate $\omega = 14.1$ (*1*), 8.8 (*2*), 5.6 (*3*), 3.4 (*4*), 2.8 (*5*), 1.7 (*6*) °C min^{-1} (DTA)

the total thermal effect $Q \approx 2.6\,kJ\,g^{-1}$, the first stage heat release is approximately equal to $0.35Q$ for a 27% weight change. Thus, the thermal effect of the dominant first-stage reaction ($Q_1 \approx 3.3\,kJ\,g^{-1}$) is much higher than that of the second-stage reaction ($Q_2 \approx 2.3\,kJ\,g^{-1}$).

The addition of Fe_2O_3 was found to result in the significant acceleration of both process stages (Fig. 9.11), which agrees with literature data [4, 5]. The dependencies of the effective activation energy on the conversion degree for both samples are presented in Fig. 9.12.

9.3 Effect of Gas-Phase Processes and Pressure on the Macrokinetics of High-Temperature Decomposition of AP-Based Composite Solid Propellants

Since model and commercial composite solid propellants are heterogeneous systems, the effect of the products of AP gasification on the decomposition mechanism is quite important. The term "gasification" is used to emphasize that at AP

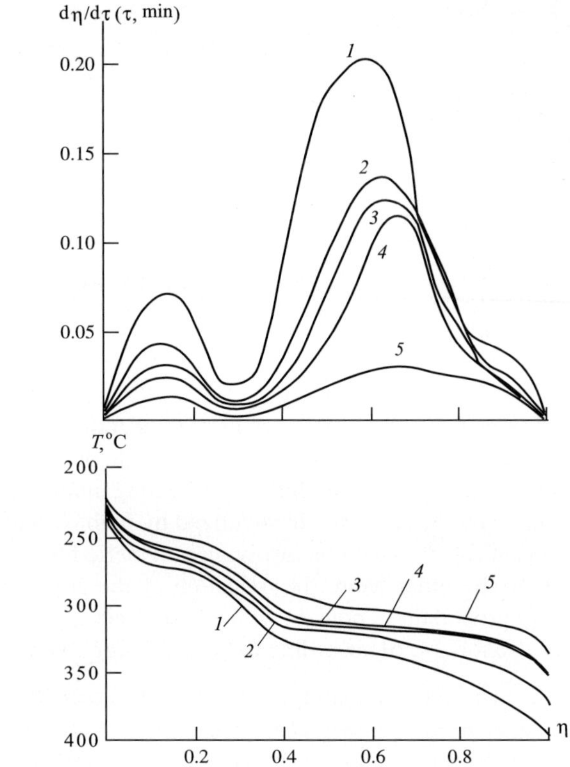

Fig. 9.10 Plots of the rate of NG-2-60-DTM (sample 1) decomposition against the conversion degree at heating rate $\omega = 5.8$ (*1*), 3.2 (*2*), 2.95 (*3*), 2.15 (*4*), 1.35 (*5*) °C min^{-1} (TGA)

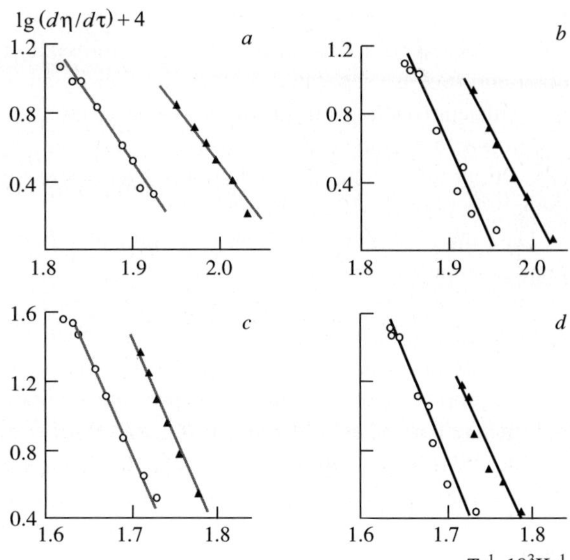

Fig. 9.11 Plots of the high-temperature decomposition rate of NG-2-60-DTM: sample 1 without the catalyst (*circles*) and sample 2 with the catalyst (0.5% Fe_2O_3) (*triangles*) at conversion degree $\eta = 0.1$ (**a**), 0.2 (**b**), 0.4 (**c**) and 0.6 (**d**)

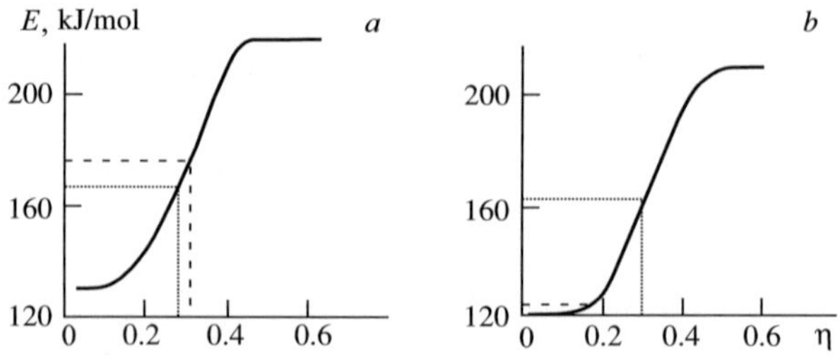

Fig. 9.12 Plots of the effective activation energy against the conversion degree for the decomposition of NG-2-60-DTM: sample 1 without the catalyst (**a**) and sample 2 with the catalyst (0.5% Fe_2O_3) (**b**)

decomposition (especially at high temperatures) two processes contribute to the formation of gaseous products: irreversible decomposition (yielding "gas") and reversible dissociative sublimation (yielding ammonia and perchloric acid: "vapor"). "Gas" and "vapor" are characterized by high oxidative abilities and can play a significant role in reactions involving polymeric fuels.

On the other hand, the oxidation of the polymeric component of a composite solid propellant in contact with an AP grain can occur without the participation of the "vapor" [4, 6] according to the following routes:

condensed compound (polymer) + solid compound (AP),
condensed compound (polymer) + gas (NO_2, O_2 and others),
condensed compound (polymer) + liquid ($HClO_4$).

In this case, a modified DTA method with thermal dilution can be applied. In this method, the decomposition of a double "sandwich"-type system (thin layers of oxidizer and fuel are separated from each other by an inert porous membrane) is studied. This approach is similar to the chemical arc used to study the high-temperature decomposition of the fuel–oxidizer system. In our experiments, the AP and polymer (PS, PE, polyurethane of DTM grade) layers were separated by a glass filter 300 μm thick with pores 5–10 μm in diameter.

Data on the kinetics of heat evolution in AP–polymer systems obtained using regular samples ("mixture," see Sect. 9.1.2) and modified samples ("sandwich") are shown in Figs. 9.13–9.16. Despite the absence of direct contact between the AP crystals and the polymer in these sandwich systems, the decomposition pattern is similar to that of the corresponding mixture systems. As the AP crystal size (d_{AP}) decreases from 200 μm to 50 μm, the second-stage decomposition rate does not change for the AP–PE, AP–PVC, AP–PAN sandwiches and slightly increases for the AP–PS and AP–DTM sandwiches. It is interesting that in the case of the AP–PE mixture, the decomposition occurs via one stage for $d_{AP} = 200$ μm and via two stages for $d_{AP} = 50$ μm, while for the AP–PE sandwich system the reaction is always characterized by two stages (Fig. 9.15). The degeneration of the first stage when coarse AP powder is used is probably associated with the physicochemical influence of the PE melt on AP decomposition.

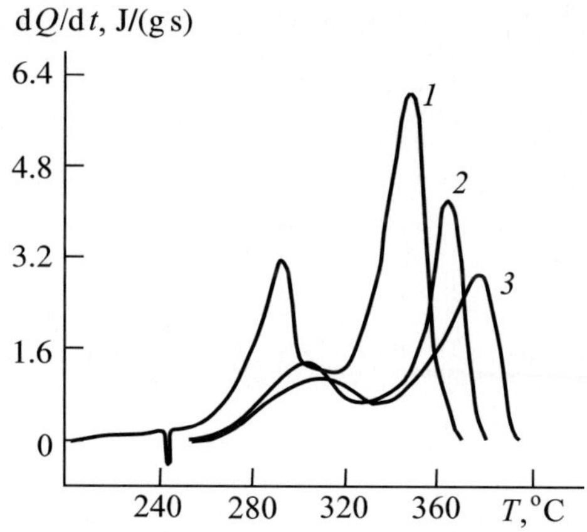

Fig. 9.13 Kinetic curves for the decomposition of heterogeneous systems: AP–DTM mixture (*1*), AP (50 μm)–DTM "sandwich" (*2*), AP (200 μm)–DTM "sandwich" (*3*)

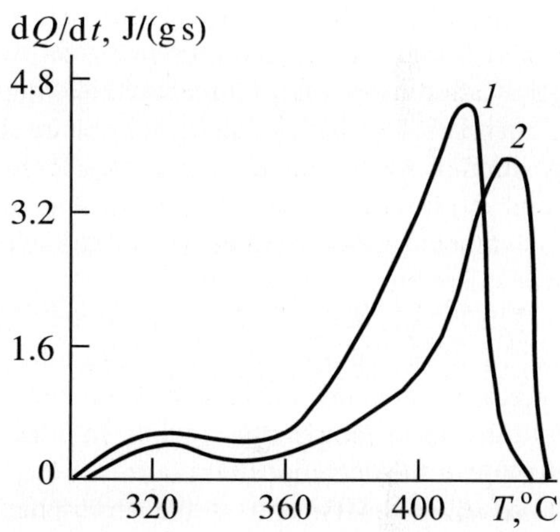

Fig. 9.14 Kinetic curves for the decomposition of heterogeneous systems: AP–PVC mixture (*1*), AP–PVC "sandwich" (*2*)

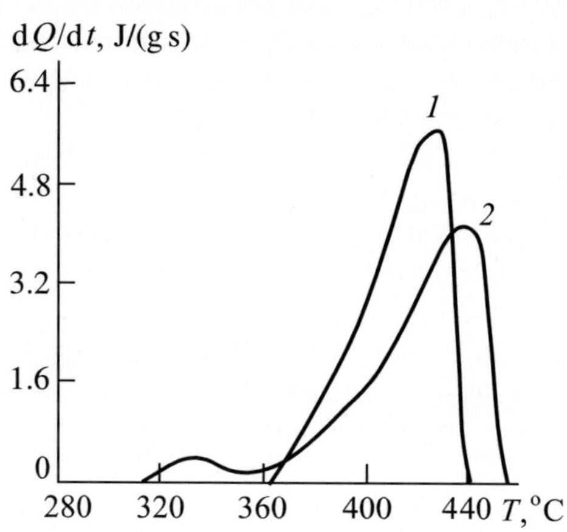

Fig. 9.15 Kinetic curves for the decomposition of heterogeneous systems: AP–PE mixture (*1*), AP–PE "sandwich" (*2*)

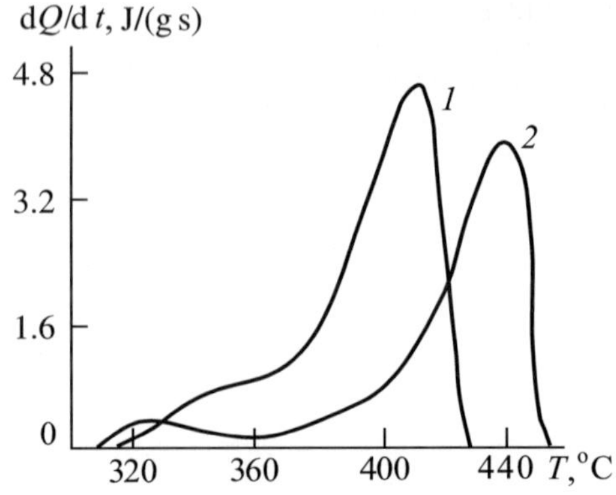

Fig. 9.16 Kinetic curves for the decomposition of heterogeneous systems: AP–PAN mixture (*1*), AP–PAN "sandwich" (*2*)

The experimental results (despite their qualitative nature) obtained for the sandwich systems indicate an important role for reactions involving the products of AP gasification in the high-temperature decomposition of AP–polymer systems.

In relation to this, the ambient pressure should significantly influence the decomposition of AP-based mixtures at high temperatures. The effect of the ambient pressure, P_{∞}, on the kinetics of the high-temperature decomposition of heterogeneous systems (the AP–PS model system and NG-2-60-DTM composite solid propellant) was studied using installation T-2.

The kinetic dependencies for the decomposition of the AP–PS mixture at $P_{\infty} = 10^2$ Pa, 10^5 Pa and 1 MPa ($\omega = 3\,\text{deg min}^{-1}$) are shown in Fig. 9.17. The pressure decrease results in a shift in the decomposition process (especially the second stage) into the lower temperature region. In other words, the lower the ambient pressure, the higher the decomposition rate.

At $P_{\infty} = 1$ MPa and at the atmospheric pressure, the second-stage reaction ($0.3 < \eta < 0.9$) is of pseudo-zero order. However, at higher pressure the process is characterized by a lower activation energy (Fig. 9.18). The expression for the rate constant is

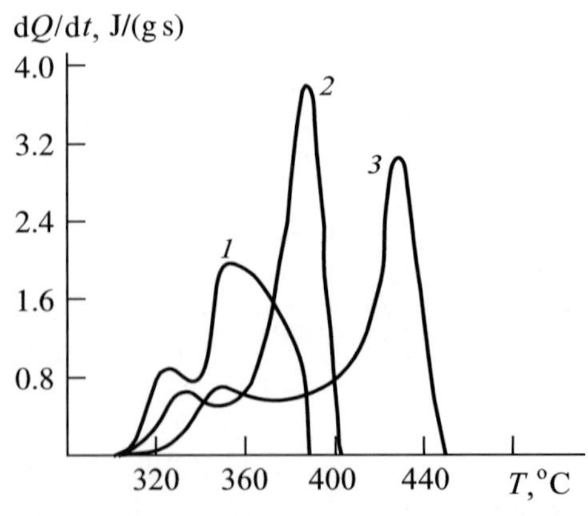

Fig. 9.17 Kinetic curves for the decomposition of the AP–PS mixture ($\omega = 3$°C min^{-1}) at $P_{\infty} = 10^2$ Pa (*1*), 10^5 Pa (*2*), 1 MPa (*3*)

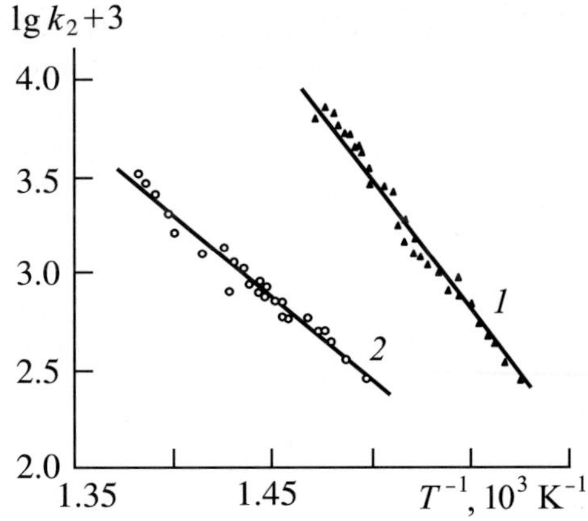

Fig. 9.18 Plots of the decomposition rate of the AP–PS mixture (second stage) against the temperature at $P_\infty = 10^5$ Pa (*1*), 1 MPa (*2*)

$$k_2 = 10^{9.8\pm0.5}\exp\left(-\frac{40900 \pm 1500}{RT}\right)\left[s^{-1}\right]. \tag{9.6}$$

The same pattern was observed for the NG-2-60-DTM material (Fig. 9.19). The influence of the ambient pressure on the first-stage process was found to be similar but much weaker.

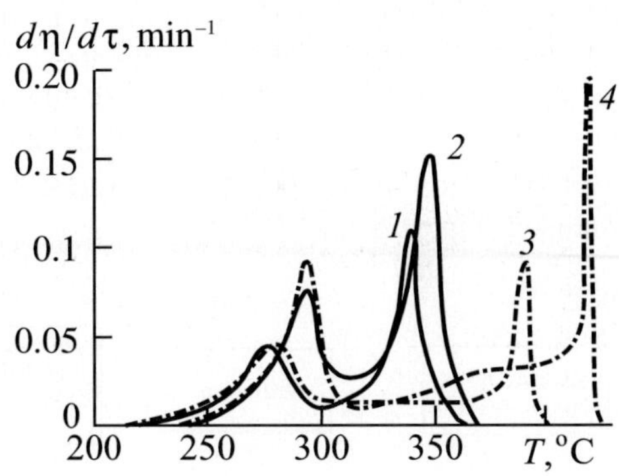

Fig. 9.19 Effect of P_∞ on the rate of NG-2-60-DTM decomposition: *1* and *2*, $P_\infty = 10^5$ Pa; *3* and *4*, $P_\infty = 1$ MPa; *1* and *3*, $\omega = 2.9\,\mathrm{deg\,min^{-1}}$; *2* and *4*, $\omega = 3.4\,\mathrm{deg\,min^{-1}}$

9.4 Mechanism of High-Temperature Decomposition of AP-Based Model Mixtures and Composite Solid Propellants

The data on the thermal decomposition of the heterogeneous condensed systems studied indicate a complex process. The decomposition of AP and most of the mixtures occurs via two stages, with the exception of the AP (coarse)–PE system, which decomposes in a single stage. The organic binder mainly influences the first stage of AP decomposition. For the AP–PS, AP–PAN and AP–PVC mixtures, the first decomposition stage ($Q_1 = 0.2$–$0.5\,\mathrm{kJ\,g^{-1}}$, $E_1 = 133$–$142\,\mathrm{kJ\,mol^{-1}}$, $T_1 = 290$–370°C) is a heterogeneous reaction involving the thermal oxidative rupture of the polymer by the products of the irreversible AP decomposition. The first-stage reaction occurs in diffusion mode with respect to the oxidant gas.

The first stage of the decomposition of NG-2-60-DTM ($Q_1 = 0.9$–$1.1\,\mathrm{kJ\,g^{-1}}$, $E_1 = 125$–$146\,\mathrm{kJ\,mol^{-1}}$, $T_1 = 220$–$290°\mathrm{C}$ at $\omega = 1.7$–$14\,\mathrm{deg\,min^{-1}}$) seems to take place in the condensed phase, since the reaction rate is not influenced by the ambient pressure (Fig. 9.19). The difference between the first-stage processes for NG-2-60-DTM and the model mixtures mentioned above is probably related to the presence of an ether group in the backbone chain of the NG-2-60-DTM binder.

A first stage characteristic of AP decomposition was observed when the oxidizer and fuel condensed phases were separated by a porous glass membrane. Therefore, the absence of the first stage during the decomposition of the AP (coarse)–PE system may be explained by the inhibitory effect of the PE melt. A similar effect of saturated hydrocarbons (hexane) on the macrokinetics of AP decomposition at low temperatures was reported for the first time in [7].

The second stage of the decomposition of the mixture (60–90%) is characterized by comparatively high activation energies: $E_2 = 217$–$263\,\mathrm{kJ\,mol^{-1}}$. These values, which are close to the activation energies of thermal destruction processes for polymers, differ significantly from those characteristic of the second stage of AP decomposition ($E_2^{\mathrm{AP}} = 133\,\mathrm{kJ\,mol^{-1}}$). However, although perceptible in the overall decomposition processes of AP–PS and AP–DTM systems, the effect of the thermal destruction of the binder was found to be negligible for the AP–PE, AP–PAN and AP–PVC mixtures (in the experiments, samples with fuel and oxidizer condensed phases separated by a porous glass membrane were used).

As a result of the qualitative analysis of the second stage of the high-temperature decomposition of AP–polymer mixtures, the following important steps can be formulated (Fig. 9.20):

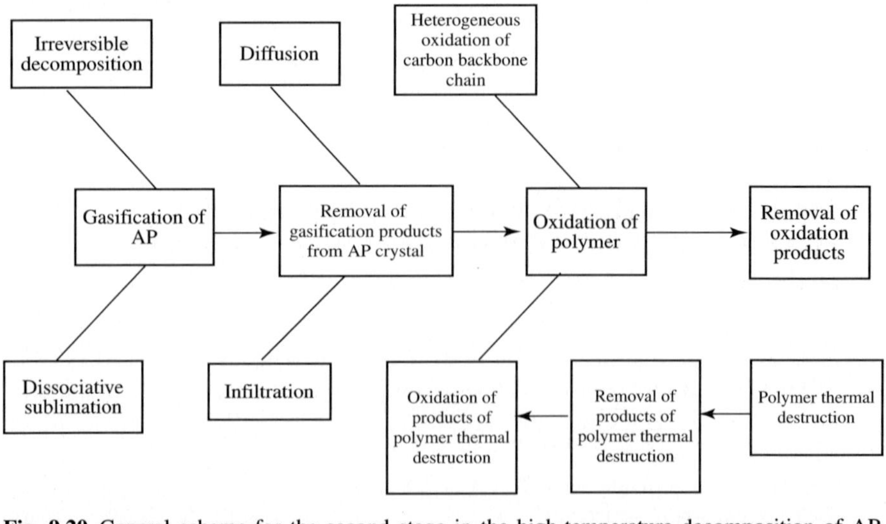

Fig. 9.20 General scheme for the second stage in the high-temperature decomposition of AP–polymer heterogeneous systems

a) Irreversible decomposition and dissociative sublimation of AP; formation of liquid and gaseous perchloric acid which partially decomposes in the pores of AP (the formation and accumulation of liquid perchloric acid in the pores of AP takes place due to the higher volatility of ammonia than perchloric acid vapor, which also seems to be related to the increase in the thermal decomposition rates of some ammonia salts, including AP, after their vacuumization); transport of two gaseous oxidizers to the polymer surface: the perchloric acid vapor and the products of its decomposition ("strong" oxidizer), and the products of the irreversible decomposition of AP ("weak" oxidizer)
b) Diffusion and infiltration of the gaseous oxidizers through the inner pores of the AP crystals to the binder surface; molecular diffusion dominates in the mass-transfer process due to the small particle size of the AP–polymer system components and the high porosity of the AP crystals
c) Thermal destruction of the polymer and reactions involving gaseous destruction products
d) Oxidation of the polymer; heterogeneous oxidation of carbonized polymer residue (carbon residue produced by the oxidation of the most reactive parts of the polymer during the first stage) and gas-phase oxidation of the products of thethermal destruction of the polymer can take place.

The rate constants of the second stage of decomposition for systems containing polymers that differ significantly in terms of their structures are quite close. This (as well as the high activation energies) can be explained by taking the effect of the volumetric evaporation accompanying AP decomposition into account. The oxidizing ability of hot perchloric acid vapor with respect to organic compounds is much higher than those of nitrogen oxides and chlorine (the primary products of the irreversible decomposition of AP). Thus one can assume that at high temperatures the observed rate of system decomposition is determined by the rate of perchloric acid vapor transport from the AP crystal volume to the crystal–fuel interface ($\dot{m}_{\mathrm{tr}}$) and by the specific area of the interface ($S_{\mathrm{sp}} \sim 1/d$). According to [8], at $P_{\mathrm{sv}} \ll P_\infty$

$$\dot{m}_{\mathrm{tr}} \sim \dot{\eta}\frac{P_{\mathrm{sv}}}{P_\infty}S_{\mathrm{sp}}, \qquad (9.7)$$

from which the activation energy of the overall process is

$$E = E_{\mathrm{AP}} + \frac{\Delta H}{2} = 129 + 121 = 250\,\mathrm{kJ\,mol^{-1}},$$

which is in good agreement with the corresponding values for E_2 (Table 9.2). Since under these conditions the decomposition kinetics are determined by the properties of AP rather than those of the polymers, the rate constants for different mixtures should be similar.

The validity of the suggested mechanism is also confirmed by the paradoxical phenomenon of the inhibition of the high-temperature decomposition of AP–polymer mixtures as the ambient pressure increases. The temperature dependencies of the rate constants for the decomposition of the AP–PS mixture at $P_\infty = 10^5$ Pa

Table 9.2 Kinetic parameters for high-temperature decomposition of AP–polymer model systems [9]

Mixture	Conversion degree range	Activation energy E_2, kJ mol^{-1}	Preexponential factor k_{02}, s^{-1}	Reaction order n	Total thermal effect Q, kJ g^{-1}
AP–PS	0.40–0.85	262 ± 10	$10^{18.8\pm0.7}$	0	2.5 ± 0.2
AP–PVC	0.30–0.80	261 ± 12	$10^{18.2\pm0.9}$	1	2.2 ± 0.2
AP–PAN	0.30–0.80	250 ± 14	$10^{18.6\pm1.1}$	1	2.3 ± 0.2
AP–PE	0.05–0.80	250 ± 11	$10^{16.9\pm1.0}$	1	2.2 ± 0.2

and 1 MPa are given in Fig. 9.18. At the higher pressure the decomposition rate is lower and the activation energy decreases from $E_2 = 263\,\mathrm{kJ\,mol^{-1}}$ (at $P_\infty = 10^5$ Pa) to $E_2 = 167\,\mathrm{kJ\,mol^{-1}}$ (at $P_\infty = 1$ MPa). The fuel is oxidized by both the gas and the vapor (perchloric acid) produced by AP decomposition. Therefore, taking into account Eq. (9.7), one can write for the overall rate of decomposition of the mixture

$$\dot{\eta}_\Sigma = \dot{\eta}_\mathrm{g} + \dot{\eta}_\mathrm{v} \approx k_\mathrm{g}\dot{m}_\mathrm{g}\left(1 + \frac{k_\mathrm{v}\mu_\mathrm{v}P_\mathrm{sv}}{k_\mathrm{g}\mu_\mathrm{g}P_\infty}\right), \tag{9.8}$$

where $\dot{\eta}_\mathrm{g}$, $\dot{\eta}_\mathrm{v}$ and k_g, k_v are the rates and the rate constants of the heterogeneous oxidation of the polymer by gas and vapor, respectively.

Since $k_\mathrm{v} \gg k_\mathrm{g}$, the kinetics of mixture decomposition are determined by the vapor formation rate, even at low P_sv/P_∞ ratios. In this case, the activation energy is close to $250\,\mathrm{kJ\,mol^{-1}}$ (assuming low activation energies for heterogeneous oxidation reactions). As P_∞ grows, the P_sv/P_∞ ratio and the second term inside the parentheses of Eq. (9.8) decrease in proportion to it. As a result, the overall rate of mixture decomposition decreases. The effective activation energy also decreases down to a limiting value of $125\,\mathrm{kJ\,mol^{-1}}$, corresponding to AP decomposition. At very low P_sv/P_∞ ratios (low temperatures), the decomposition macrokinetics for the AP–polymer mixtures can become very similar to the kinetics of AP decomposition. This was observed for the low-temperature decomposition (first stage) of the AP–PS mixture at $P_\infty = 10^5$ Pa [10].

9.5 Relationship Between the Kinetics of High-Temperature Decomposition of AP-Based Composite Solid Propellants and Their Ignition and Subatmospheric Burning Parameters

This section is devoted to the kinetics and macrokinetics of the high-temperature decomposition of AP-based composite solid propellants. A detailed study of the relationship between the kinetics of reactions in the condensed phase and burning patterns is not provided here. This aspect has been already considered in a number of works (for example, see [6, 11, 12, 13]). Due to the absence of kinetic data for temperatures close to the surface temperature of the ignited or burning fuel, a quantitative analysis of the contribution of condensed-phase reactions to the overall process was previously not possible. However, this can now be achieved using the kinetic data presented in this chapter and in Chap. 5, although it is worth noting that only cases where the reacting mixture can be assumed to be pseudohomogeneous can be considered. In other words, the reacting layer thickness should significantly exceed the characteristic heterogeneity scale (in this case, the AP crystal size).

Practically speaking, this means that the obtained kinetic parameters can be used to analyze processes characterized by an ignition delay (induction) period of more than 0.5 s. Such an induction period is characteristic of solid-propellant rocket charges undergoing aerodynamic heating for example. The obtained kinetic constants can also be used to quantitatively analyze the role of condensed-phase reactions in AP-based composite solid propellant burning at subatmospheric pressures. At $P \leq 0.1$ MPa, the reaction rates for most AP-based composite solid propellants are quite low. Therefore the reaction zone can be treated as being homogeneous, since its thickness significantly exceeds the AP grain size.

First, let us compare kinetic data obtained by DTA and the method of ignition using a hot metal block (Sect. 5.4) for two samples of NG-2-60-DTM commercial polyurethane composite solid propellant (with and without the combustion catalyst, Fe_2O_3).

The values of Qk_0 and E_{ign} were obtained by experiments on ignition in the temperature range of 350–430°C. In this case, a term characterizing the conversion degree for the pre-ignition time was neglected in Eq. (5.5)

$$\tau_{\text{ign}} = (1 + 1.6\Theta_{0\infty} + 0.2\Theta_{0\infty}^2)(1 + 8\Theta_{0\infty}\gamma^{3/2})(1 + \beta).$$

By taking the term $8\Theta_{0\infty}\gamma^{3/2}$, which includes the thermal effect Q ($\gamma = RT_S^2 c/QE$; Q is determined by thermography), into account, refined data for E'_{ign} were obtained (Table. 9.3).

The difference between E'_{ign} and E_{ign} was found to be approximately 3 kJ mol^{-1}, which is within the experimental error.

Table 9.3 Determination of the conversion degree η_{ign} on the surface of the composite solid propellant at its ignition (comparison of data obtained by the ignition method and DTA)

Composition	Ignition on a metal block				DTA with thermal dilution		
	Data from Chap. 5		Refined data				
	E_{ign}, kJ mol^{-1}	$\lg Qk_0$, J g^{-1} s^{-1}	E'_{ign}, kJ mol^{-1}	$\lg Qk_0$, J g^{-1} s^{-1}	$E = E'_{\text{ign}}$, kJ mol^{-1}	η_{ign}	$\lg Qk_0$, J g^{-1} s^{-1}
AP–polyurethane	178	16.4	182	16.8	182	0.3	16.6
AP–polyurethane + Fe_2O_3	129	13.2	131	13.6	131	0.17	13.4

The values of η_{ign} corresponding to the pre-ignition period were estimated using the $E_{eff}(\eta)$ dependence (characteristic of a two-stage process), as obtained by thermography. This approach is usually used to analyze kinetic data for multistage processes and when the compositions of the intermediate and final reaction products are unknown. In this case, formally describing the kinetics using the effective rate, conversion degree and activation energy greatly simplifies the analysis of the high-temperature decomposition of AP–polymer systems. In Table 9.3, three values for Qk_0 are given: data from Chap. 5, refined data, and data calculated from DTA data using the dependence of the heat release rate on the conversion degree for $\eta = \eta_{ign}$ at $T = T_S = \text{const}$. The obtained η_{ign} (= 0.2–0.3) indicates that significant burning-out occurs in the reaction zone of a heterogeneous system during the induction period.

The kinetic constants for decomposition calculated from experimental data on combustion at low pressures were compared with the characteristics obtained by the thermographic method for both polyurethane-based composite solid propellants and AP–PS and AP–PE mixtures. The burning rates of these materials at pressures of $0.02\,\text{MPa} < P \leq 0.1\,\text{MPa}$ were measured by a standard method using a constant-pressure bomb.

The experimental data were approximated using the expression

$$u = AP^{\nu}. \tag{9.9}$$

The values of A and ν are given in Table 9.4.

The kinetic constants were calculated using the model for the burning of composite solid propellants developed by Waesche and Wenograd for the first time [6]. A similar model for homogeneous propellants was used by Zel'dovich [16]. According to the approach used, the rate of burning at $P \leq 10^5$ Pa is assumed to be determined by condensed-phase reactions, while the surface temperature of the burning fuel is identical to the temperature of the equilibrium dissociative sublimation of AP. As was stated earlier, condensed-phase reactions should play a very important role in the burning of composite solid propellants, at least at $P \leq 10^5$ Pa. The existence of the relationship between the surface temperature of the burning composite solid propellant and the temperature of equilibrium sublimation of AP was

Table 9.4 Comparison of the combustion parameters and kinetic characteristics of high-temperature decomposition of AP-based composite solid propellants at $P = 0.1$ MPa ($T_S = 545$°C)

Composition	Combustion				DTA		
	A	ν	E,* kJ mol^{-1}	$\lg Qk(T_S)$ J g^{-1} s^{-1}	E, kJ mol^{-1}	η_S	$\lg Qk(T_S)$ J g^{-1} s^{-1}
NG-2-60-DTM	1.05 ± 0.02	0.72 ± 0.04	174	4.88	174	0.27	4.7
NG-2-60-DTM	1.30 ± 0.02	0.70 ± 0.04	169	5.06	169	0.23	4.9
AP–PS	0.79 ± 0.02	0.93 ± 0.03	225	4.75	225	0.33	3.9
AP–PE	0.58 ± 0.01	0.88 ± 0.03	213	4.45	250	0.35	4.0

* See expression (9.11).

proved experimentally [4, 14, 15]. However, the correctness of these models can be quantitatively confirmed only through the use of kinetic data obtained for high temperatures.

In general, the surface temperature of a burning volatile energetic material (a liquid explosive or some other material), T_S, differs from the sublimation (evaporation) temperature, T_{sub}, at thermodynamic equilibrium due to the escape of the sublimate and its dilution with the burning products [16]. However, this difference is quite small, and the conditions on the surface of the sublimating AP can be considered as being in thermodynamic equilibrium to a first approximation. Besides, the characteristic time required for the spatial equalization of the temperature in the direction perpendicular to the front of a comparatively slow-burning composite solid propellant in vacuum is quite short. Therefore, the temperatures of all of the parts of a burning composite solid propellant can be considered to be identical, constant and close to the temperature of AP equilibrium sublimation: $T_S = T_{sub}(P)$. Thus T_S was calculated using the expression [17]

$$\lg P = 10.56 - \frac{6283.7}{T_S}. \tag{9.10}$$

For the burning model used,

$$v = \frac{mE}{2\Delta H}, \tag{9.11}$$

where $m = 2$ (AP dissociates yielding two gases), and so Eq. (9.11) was used to estimate the effective activation energy of decomposition (Table 9.4).

The heat evolution rate ($\dot{q} = Qk(T_S)$) was determined using the Zel'dovich formula [16]

$$u = \sqrt{\frac{2\lambda R T_S^2 Q k_0}{\rho E c^2 (T_S - T_\infty)}} \exp\left(-\frac{E}{RT_S}\right), \tag{9.12}$$

where $\lambda = 2.6 \times 10^{-3}\,\mathrm{J\,cm^{-1}\,s^{-1}\,deg^{-1}}$ and $c = 1.7\,\mathrm{J\,g^{-1}\,deg^{-1}}$.

The conversion degree on the burning surface was determined from the $E(\eta)$ dependencies, as obtained by thermography (this method was used to estimate the burning-out for the ignition delay period). The heat evolution rate, $\dot{q}$, at temperature T_S and the estimated conversion degree η_S was calculated using the relationship $\dot{q} = \dot{\eta} Q$ and the experimentally determined $\dot{\eta}(\eta)$ dependencies for $T = T_S$.

The values of $\lg Qk(T_S)$ determined by the two methods are in good agreement. The conversion degree values fall within quite a narrow range ($\eta_S = 0.25$–0.35) and are close to those calculated from the heat-balance equation ($\eta_S = 0.33$–0.35).

The effective activation energy of the high-temperature decomposition of the AP–PE mixture was found from the DTA data to be constant over the entire conversion degree range. Therefore the conversion degree on the burning surface of this mixture, η_S, was estimated using the heat-balance equation. The effective activation energy obtained by combustion and thermographic methods ($208 < E_{eff} < 250\,\mathrm{kJ\,mol^{-1}}$) is in excellent agreement with that found by the method of AP–PE mixture ignition on a hot block ($E_{ign} = 233\,\mathrm{kJ\,mol^{-1}}$).

Such a good agreement between the kinetic characteristics determined by three methods indicates that the thermal decomposition mechanism for AP-based composite solid propellants is the same over a wide temperature range.

References

1. Grassi N (1959) Chemistry of polymer destruction. Inostrannaya Literatura, Moscow
2. Madorsky SL (1964) Thermal degradation of organic polymers. Wiley, New York
3. Mal'kov YuI, Benin AI, Kossoy AA (1974) Vysokomol Soedin 16:2738
4. Korobeynichev OP (1971) (ed) Mechanism, kinetics and catalysis of thermal decomposition and combustion of ammonium perchlorate. Nauka, Novosibirsk
5. Manelis GB, Nazin GM, Rubtsov YI, Strunin VA (2003) Thermal decomposition and combustion of explosives and propellants. Taylor & Francis, New York
6. Waesche RHW, Wenograd J (1967) 2nd Solid Propulsion Conf, New York, USA, June 1967
7. Andreev KK (1967) Theory of explosives. Visshaya Shkola, Moscow
8. Merzhanov AG, Slutsker BM, Shteinberg AS (1968) Combust Explos Shock Waves 4:312
9. Merzhanov AG, Shteinberg AS, Goncharov EP (1973) Combust Explos Shock Waves 9:157
10. Goncharov EP, Merzhanov AG, Shteinberg AS (1972) In: Stesik LN (ed) Combustion and explosion. Nauka, Moscow, p 765
11. Bakhman NN, Belyaev AF (1967) Combustion of heterogeneous condensed systems. Nauka, Moscow
12. Manelis GB, Strunin VA (1972) In: Stesik LN (ed) Combustion and explosion. Nauka, Moscow, p 53
13. Manelis GB, Strunin VA (1971) Combust Flame 17:69
14. Powling J, Smith WA (1962) Combust Flame 6:173
15. Powling J, Smith WA (1963) Combust Flame 7:269
16. Zel'dovich JaB (1942) Zh Eksp Teor Fiz 12:498
17. Jnami S, Rosser W, Weis NM (1963) J Phys Chem 67:1077

Chapter 10
Theory of Decomposition Macrokinetics and Thermal Explosion of Volatile Explosives

Abstract A theory for the thermal explosion of volatile energetic materials (liquid explosives widely used in rocket engineering, such as highly concentrated H_2O_2, hydrazine, ammonium perchlorate and nitrate, and others) is considered. The irreversible decomposition of the condensed phases of these materials are accompanied by reagent evaporation (in the case of a solid compound, sublimation) and the saturation of the gaseous reaction product with reagent vapor. This effect should be taken into consideration when analyzing reaction-related processes, particularly heat evolution. The macrokinetics of these processes are discussed. Thermal explosion of volatile energetic material is analyzed in terms of the classical theory. The thermal explosion diagram for volatile energetic material (the Semenov diagram) is shown to differ significantly from those of gases and nonvolatile condensed energetic materials. Expressions for the critical conditions for the thermal explosion of volatile energetic materials that can be used for engineering calculations are obtained. The correctness of the proposed theory of thermal explosion is confirmed by experimental data on the kinetics of heat evolution and thermal explosion of dinitroxydiethylnitramine.

10.1 Introduction

Results from detailed theoretical and experimental studies of thermal decomposition and explosion of nonvolatile explosives can be found in the literature (see for example [1]). The decomposition of volatile explosives have been investigated without taking the effect of evaporation into account quantitatively (the important role of this process was emphasized in [2, 3], and some approaches to solving this problem can be found in [4]).

The influence of evaporation on the macrokinetics of thermal decompositions of volatile explosives in open systems for zero- and first-order reactions is analyzed in this chapter.

When evaporation accompanies volumetric thermal decomposition of a liquid, the phase transition occurs on the inner surfaces of gas bubbles (in volume) rather than on the liquid surface (plane). Due to high mass- and heat-transfer parameter values, the gas in the growing bubbles is continuously saturated with vapor of the liquid (or partially decomposed liquid). As opposed to evaporation from the liquid surface, the process rate is determined by the volumetric reaction kinetics and by the humidity of the gas produced rather than by outward mass- and heat-transfer parameters. In other words, the evaporation is a volumetric process. The chemical reaction rate is significantly influenced by the evaporation since it occurs with heat absorption and results in a reduction in the amount of reagent.

The situation where evaporation from the liquid surface prevails over the volumetric process (for example, in very slow reactions in a thin liquid layer) is not considered here, since it is not characteristic of explosion.

10.2 Kinetics of Thermal Decomposition of Volatile Explosives

Let us consider the kinetics of the reduction in the weight of a liquid under the conditions of a thermogravimetric experiment at constant temperature. Under quasi-stationary conditions, the total rate of the weight reduction ($-dm/dt$) can be expressed as the sum of the gas formation and evaporation rates (dm_g/dt and dm_v/dt, respectively):

$$-\frac{dm}{dt}=\frac{dm_g}{dt}+\frac{dm_v}{dt},\qquad -\frac{dm}{dt}=\frac{dm_g}{dt}\left(1+\frac{dm_v}{dm_g}\right). \tag{10.1}$$

In standard kinetic experiments the process rate is considerably low, while the mass- and heat-transfer rates inside the bubbles are extremely high due to their small size. Taking these circumstances into account, one can assume the existence of thermodynamic equilibrium inside a bubble. In this case,

$$\frac{dm_v}{dm_g}=\frac{\mu_v P_v}{\mu_g P_g}.$$

Taking into consideration that $P_v+P_g=P_\infty$ and $P_v=P_0\exp(-L/RT)$, one obtains

$$\frac{dm_v}{dm_g}=\frac{\mu_v}{\mu_g}\frac{P_0\exp(-L/RT)}{P_\infty-P_0\exp(-L/RT)}, \tag{10.2}$$

where T is the liquid temperature, P_∞ is the pressure in the gas bubble (equal to the ambient pressure), L is the latent evaporation heat, P_0 is the preexponential factor in the expression for the saturated vapor pressure of the original liquid, μ_v and μ_g are the molecular weights of original liquid vapor and the gas produced, respectively, and R is the gas constant.

Substituting Eq. (10.2) into Eq. (10.1), one obtains

$$-\frac{dm}{dt}=\frac{dm_g}{dt}\left[1+\frac{\mu_v}{\mu_g}\frac{P_0\exp(-L/RT)}{P_\infty-P_0\exp(-L/RT)}\right]. \tag{10.3}$$

The gas formation rate can be expressed by using the relationship for a first-order reaction

$$\frac{dm_g}{dt} = k(T)m. \tag{10.4}$$

Using this, one obtains for the effective mass reduction rate constant

$$k_{\mathrm{eff}} = k(T)\left[1 + \frac{\mu_v}{\mu_g}\frac{P_0 \exp(-L/RT)}{P_\infty - P_0 \exp(-L/RT)}\right]. \tag{10.5}$$

Now let us analyze the kinetics of gas evolution in this system using the same approach. The heat balance can be expressed as the sum of the heat released at decomposition (gas formation) and absorbed during evaporation:

$$\frac{dq}{dt} = Q_g\frac{dm_g}{dt} - L\frac{dm_v}{dt},$$

where Q_g is the thermal effect per gram of gas. Using mathematical manipulations identical to Eqs. (10.1)–(10.4), one obtains an expression for the rate of heat evolution:

$$\frac{dq}{dt} = Q_g\frac{dm_g}{dt}\left[1 - \frac{L\mu_v}{Q_g\mu_g}\frac{P_0 \exp(-L/RT)}{P_\infty - P_0 \exp(-L/RT)}\right]. \tag{10.6}$$

If the overall decomposition reaction can be written as

$$aA \text{ (liquid)} \rightarrow bB \text{ (gas)} + cC \text{ (liquid or solid)},$$

then Eq. (10.6) can be transformed identically, yielding

$$\frac{dq}{dt} = Q\frac{dm_{\mathrm{lqd}}}{dt}\left[1 - \frac{Lb}{Qa}\frac{P_0 \exp(-L/RT)}{P_\infty - P_0 \exp(-L/RT)}\right], \tag{10.7}$$

where dm_{lqd}/dt and Q refer to the decomposing liquid A.

In this case, the measured value of the thermal effect of the process, Q_{eff}, depends on the true value of the thermal effect, Q, and on the experimental temperature and pressure (for a constant reaction mechanism):

$$Q_{\mathrm{eff}} = Q\left[1 - \frac{L\mu_v}{Q\mu_g}\frac{P_0 \exp(-L/RT)}{P_\infty - P_0 \exp(-L/RT)}\right]. \tag{10.8}$$

This dependence is qualitatively illustrated in Fig. 10.1. For a certain temperature, T_{ad} (independent of the reaction kinetics and dependent on P_∞), it can be determined from the equation

$$1 - \frac{L\mu_v}{Q\mu_g}\frac{P_0 \exp(-L/RT_{\mathrm{ad}})}{P_\infty - P_0 \exp(-L/RT_{\mathrm{ad}})} = 0,$$

$Q_{\mathrm{eff}} = 0$. In other words, the process becomes thermoneutral.

The conditions for which the influence of volumetric evaporation on the decomposition rate can be neglected can be obtained from Eqs. (10.5) and (10.6). Thus, one obtains for the gravimetric experiment:

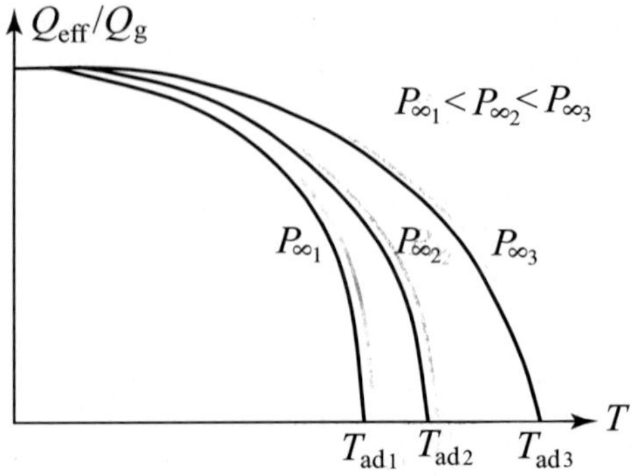

Fig. 10.1 Plot of the normalized effect Q_{eff}/Q_g against the experimental temperature and the ambient pressure

$$\frac{\mu_v}{\mu_g}\frac{P_0\exp(-L/RT)}{P_\infty - P_0\exp(-L/RT)} << 1 \tag{10.8a}$$

and for the calorimetric experiment:

$$\frac{L\mu_v}{Q\mu_g}\frac{P_0\exp(-L/RT)}{P_\infty - P_0\exp(-L/RT)} << 1. \tag{10.8b}$$

Therefore, in order to adequately study the kinetics for volatile compounds in open systems using the described approaches, the role of evaporation should be evaluated. One can use Eqs. (10.5) and (10.8) to do this.

10.3 Thermal Explosion of a Volatile Explosive

Let us consider the thermal explosion of a liquid volatile explosive. In the general case, a reaction occurring under nonisothermal conditions is accompanied by self-heating. Since the analysis of a thermal explosion in a liquid medium under complex nonisothermal conditions (due to free convection, mixing caused by the motion of bubbles, etc.) is very difficult, let us consider a system characterized by a uniform instantaneous temperature. In other words, let us analyze a thermally gradientless process that can take place at low coefficients of outward heat emission or during active mixing of the liquid. This approach to the analysis of thermal explosion has recently become quite popular due to the active development of experimental methods for studying thermally gradientless processes (see Chaps. 6–9). Taking Eq. (10.7) into account, one can write a nonstationary heat-balance equation for a zero-order reaction,

$$c\rho\frac{dT}{dt} = Qk_0\exp\left(-\frac{E}{RT}\right)\left[1 - \frac{Lb}{Qa}\frac{P_0\exp(-L/RT)}{P_\infty - P_0\exp(-L/RT)}\right] - \frac{\alpha S}{V}(T - T_0), \tag{10.9}$$

where α is the heat emission coefficient, S is the heat-exchange surface area, ρ, c, V are the density, heat capacity and volume of the liquid respectively, and T_0 is the ambient (thermostat) temperature.

The first term on the right hand side of Eq. (10.9) referring to the kinetics of heat evolution is quite complex. As the temperature grows, its value first increases (not influenced by evaporation) and reaches its maximum value, and then it decreases due to evaporation-related heat absorption. Equation (10.9) can be rewritten as

$$c\rho\frac{dT}{dt} = Qk_0\exp\left(-\frac{E}{\mathrm{R}T}\right) - Lk_0\exp\left(-\frac{E}{\mathrm{R}T}\right)\frac{b}{a}\frac{P_0\exp(-L/\mathrm{R}T)}{P_\infty - P_0\exp(-L/\mathrm{R}T)} - \frac{\alpha S}{V}(T-T_0)$$

(I) (II) (III) (IV)

so that it can be considered a heat-balance equation for a process with volumetric (term III) and surface (through the vessel walls, term IV) heat emission, similar to chain termination in the volume and on the walls in the case of an explosive chain reaction.

Let us consider possible stationary modes and critical conditions for this process. The best way to visualize the analysis is to use the Semenov diagram for the case under discussion (Fig. 10.2). The curves corresponding to heat evolution for the process with (2) and without (*1*) volumetric evaporation coincide at low temperatures (when the liquid vapor pressure is much lower than the ambient pressure). In this case, the limiting explosion temperature, T^*, exceeds the temperature characteristic of the process without evaporation, T_i^*. One can see from Fig. 10.2 that, if all other conditions are the same, the critical heat emission, $(\alpha S/V)^*$, corresponding to the explosion initiation limit for a volatile liquid is lower than that for a nonvolatile compound (characterized by the absence of volumetric heat absorption). The process characteristics corresponding to the region below the explosion initiation limit

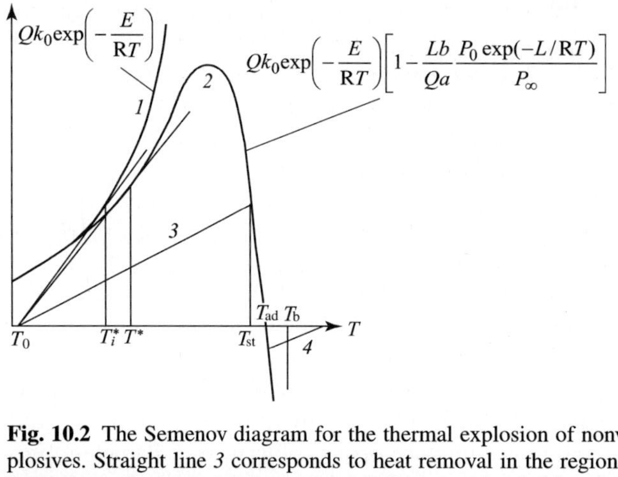

Fig. 10.2 The Semenov diagram for the thermal explosion of nonvolatile (*1*) and volatile (*2*) explosives. Straight line *3* corresponds to heat removal in the region above the explosion initiation limit, determined only by the condensed-phase reaction (without taking evaporation into account)

(the point at which the heat emission curve crosses the heat evolution curve below the inflection point) are similar to the characteristics of the process that occurs without volumetric evaporation (in the latter case the stationary heating is slightly higher). The behavior of the liquid under conditions corresponding to the region above the explosion initiation limit is of particular interest.

It is known that the temperature of the nonevaporating substance in the region above the explosion initiation limit grows rapidly (after an induction period) during thermal explosion, up to quite high values close to the adiabatic combustion (explosion) temperature. The behavior of a volatile liquid explosive is completely different. If the heat emission conditions correspond to the region above the explosion initiation limit (Fig. 10.2, line 3), a new stationary mode of the process occurs due to the specific character of the heat evolution function. This mode is characterized by a high warm-up $\Delta T_{\text{st}} = T_{\text{st}} - T_0$, which significantly exceeds the maximum warm-up $(\text{R}T_{0cr}^2/E)$ determined via the classical Semenov theory for gases and nonvolatile condensed systems.

In this case, the temperature of the liquid, T_{st}, is determined by a combination of kinetic constants, volumetric evaporation parameters and heat-transfer characteristics, and can be found by computationally solving (10.9). The upper limit of the temperature of the liquid is T_{ad}, which corresponds to the adiabatic process $\alpha S/V = 0$ (Fig. 10.1). By placing the term in square brackets in expression (10.1) equal to zero and introducing the theoretical liquid boiling point, T_{b}, determined from

$$P_\infty = P_0 \exp\left(-\frac{L}{\text{R}T_{\text{b}}}\right),$$

one obtains

$$T_{\text{ad}} = T_{\text{b}} - \frac{\text{R}T_{\text{b}}^2}{L}\frac{\ln(1+Lb/Qa)}{1+(\text{R}T_{\text{b}}/L)\ln(1+Lb/Qa)}. \tag{10.10}$$

Taking into account that, as a rule,

$$\frac{\text{R}T_{\text{b}}}{L}\ln\left(1+\frac{Lb}{Qa}\right) << 1,$$

then expression (10.10) can be simplified to

$$T_{\text{ad}} = T_{\text{k}} - \frac{\text{R}T_{\text{b}}^2}{L}\ln(1+\frac{Lb}{Qa}). \tag{10.11}$$

One can see from Eqs. (10.10) and (10.11) and Fig. 10.1 that T_{ad} increases as the ambient pressure increases and always remains below the boiling point.

At T_{st} the process occurs vigorously and is characterized by significant heat evolution, resulting in active vapor formation. In experimental studies this process is interpreted as a flash. The flash can be accompanied by a bright glow in the case of self-inflammation of the vapor–gas mixture or it can occur as a flameless process. Since it is a matter of special interest, the analysis of the critical conditions for vapor–gas mixture self-inflammation is not considered here. In experiments on thermal explosions of Tetryl and DINA (see Sect. 10.5), both flameless flashes and flashes with glow were observed. In the case of insignificant evaporation at relatively low temperatures T_0, the transition from low degrees of warm-up ($\sim RT_0^2/E$) to high ones ($\Delta T_{st} = T_{st} - T_0$) is determined by the condensed-phase reaction. The critical conditions for this process can be obtained by using standard expressions from thermal explosion theory that does not take the explosive volatility into account. It is interesting that, in the case of a flameless process taking place in the region above the explosion initiation limit, the stationary warm-up, ΔT_{st}, decreases with increasing T_0 (at $\alpha S/V = \text{const}$). As soon as T_0 reaches T_{ad}, ΔT changes sign. Upon increasing T_0 still further, the process occurs with underheating, and the temperature of the liquid approaches the theoretical boiling point.

10.4 Estimation of the Critical Conditions for Thermal Explosion of a Volatile Explosive

Let us consider the critical conditions for the thermal explosion of a liquid volatile explosive. If at T_0 the pressure of the original liquid vapor is significantly lower than the ambient pressure:

$$P_{sv}(T_0) << P_\infty, \tag{10.12}$$

then the explosion limit maximum warm-up ($\Delta T^* = T^* - T_0$) can be estimated by using a simplified expression for the heat evolution rate,

$$Qk_0 \exp\left(-\frac{E}{RT}\right)\left[1 - \frac{L}{Q}\frac{b}{a}\frac{P_0 \exp(-L/RT)}{P_\infty - P_0 \exp(-L/RT)}\right]$$

$$\approx Qk_0 \exp\left(-\frac{E}{RT}\right)\left[1 - \frac{L}{Q}\frac{b}{a}\frac{P_0 \exp(-L/RT)}{P_\infty}\right].$$

After performing the corresponding substitution in Eq. (10.9), differentiating both parts, and solving the set of two equations using the Semenov approach (equality of the heat emission and heat evolution functions and their derivatives at the contact point for the heat evolution and heat emission curves), one obtains

$$\Delta T^* = T^* - T_0 = \frac{RT^{*2}}{E}\left[1 - \frac{L^2}{QE}\frac{b}{a}\frac{P_0 \exp(-L/RT^*)}{P_\infty}\right]^{-1}.$$

After transforming this equation using the Frank–Kamenetsky approach (expansion of $\exp(-L/RT)$ around $T = T_0 + RT_0^2/E$), one obtains

$$\Delta T^* = \frac{RT_0^2}{E}\left[1 + \frac{L^2 b \exp(L/E)}{QEa \exp(-L/RT_b)} \exp\left(-\frac{L}{RT_0}\right)\right]. \tag{10.13}$$

Taking into account condition (10.12), one can obtain an expression for the critical condition from Eq. (10.13):

$$\frac{VQk_0\exp(-E/RT_0)}{\alpha S}\frac{eE}{RT_0^2}\left[1-\frac{Lb}{Qa}\left(1+\frac{L}{E}\right)\frac{\exp(-L/RT_0)}{\exp(-L/RT_b)}\exp\frac{L}{E}\right]=1.$$

This expression differs from the Semenov critical condition by the term in the square brackets on the left hand side. According to this expression, the critical conditions (in the general case) for a volatile explosive can be attained at lower heat removal (emission) rates due to partial heat removal by volumetric evaporation. On the other hand, if the liquid vapor pressure is much lower than the ambient pressure or the thermal effect of the reaction is much higher than the phase transition heat, that is

$$\exp\left(\frac{L}{E}\right)\frac{Lb}{Qa}\left(1+\frac{L}{E}\right)\frac{\exp(-L/RT_0)}{\exp(-L/RT_b)}<<1,$$

the critical condition obtained becomes identical to the Semenov one (valid in the absence of volumetric evaporation).

Since $\exp(L/E)(Lb/Qa)(1+L/E)\approx 1$ for many practically interesting systems, the classical theory, which does not take volumetric evaporation into account, can be used to estimate the critical parameters for thermal explosion provided $P_{sv}(T_0)\ll P_\infty$.

Now let us consider the critical conditions for thermal explosion (flameless flash) in a liquid explosive in a more general form.

Introducing the generalized function of heat evolution

$$F(T)=\frac{\alpha S}{V}(T-T_0),$$

where

$$F(T)=Q\rho k_0\exp\left(-\frac{E}{RT}\right)\left\{1-\frac{LbP_0\exp(-L/RT)}{Qa\left[P_\infty-P_0\exp(-L/RT)\right]}\right\},$$

one can rewrite the heat-balance equation for stationary conditions.

Using the Semenov method, one can obtain a set of equations that determine the critical conditions

$$F(T^*)=\frac{\alpha S}{V}(T^*-T_0),$$

$$F'(T^*)=\frac{\alpha S}{V},$$

where T^* is the temperature of the liquid under the conditions of the explosion initiation limit.

These equations implicitly connect the following process parameters in the vicinity of the explosion limit: the ambient temperature, T_0, the reduced heat emission, $\alpha S/V$, and the temperature of the liquid, T^*. The analytical expression for the warm-up and heat transfer in the vicinity of the explosion initiation limit cannot be obtained in its explicit form due to the significant complexity of $F(T)$. However, they can be written in a simple parametric form:

$$T_{0\text{cr}} = T^* - \frac{F(T^*)}{F'(T^*)},$$

$$\left(\frac{\alpha S}{V}\right)_{\text{cr}} = F'(T^*).$$

Due to the inflection of the $F(T)$ curve at T^*_{lim} determined by the equation $F''(T) = 0$, the system can attain the critical conditions for thermal explosion only if

$$\frac{\alpha S}{V} < \left(\frac{\alpha S}{V}\right)^{\text{cr}}_{\text{lim}},$$

$$\left(\frac{\alpha S}{V}\right)^{\text{cr}}_{\text{lim}} = F'(T^*_{\text{lim}}).$$

At $\frac{\alpha S}{V} > \left(\frac{\alpha S}{V}\right)^{\text{cr}}_{\text{lim}}$ only a single value of the liquid temperature corresponds to any value of T_0. The same restriction is applied to the ambient temperature. The critical transition is feasible only if

$$T_0 < T^{\text{cr}}_{0\text{lim}} \text{ and } T^{\text{cr}}_{0\text{lim}} = T^*_{\text{lim}} - \frac{F(T^*_{\text{lim}})}{F'(T^*_{\text{lim}})}.$$

This peculiar dependence of the stationary warm-up on the process parameters is qualitatively illustrated by the diagram for a model liquid presented in Fig. 10.3. The following inequalities correspond to curves 1–7:

$$0 = \left(\frac{\alpha S}{V}\right)_1 < \left(\frac{\alpha S}{V}\right)_2 < \left(\frac{\alpha S}{V}\right)_3 < \left(\frac{\alpha S}{V}\right)_4$$
$$= \left(\frac{\alpha S}{V}\right)^{\text{cr}}_{\text{lim}} < \left(\frac{\alpha S}{V}\right)_5 < \left(\frac{\alpha S}{V}\right)_6 < \left(\frac{\alpha S}{V}\right)_7 \to \infty.$$

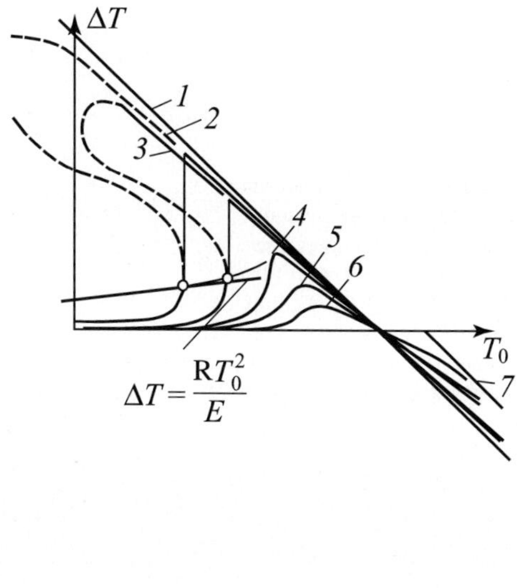

Fig. 10.3 Plot of the liquid warm-up against the ambient temperature at constant ambient pressure (see explanation in the text)

10.5 Verification of Theoretical Results Using Data on the Thermal Explosion of DINA

The main theoretical conclusions and results were qualitatively verified using data for the decomposition and thermal explosion of dinitroxydiethylnitramine (DINA, a typical secondary "volatile" explosive). DINA is one of the most convenient model compounds for studying thermal explosions of liquids due to the pseudo-zero-order reaction of the initial decomposition stage (due to the relatively small amount of condensed decomposition products [5]). The decomposition of DINA was also investigated in earlier works [6, 7].

10.5.1 Experimental Technique

Experimental studies of the macrokinetics of the thermal decomposition and explosion of DINA were carried out using an absolute (nondifferential) thermograph with variable pressure in the working chamber (Fig. 10.4) at temperatures of 130–180°C (experiments on thermal decompositions and explosions of liquids were performed at constant temperatures (T_0) in a thermostat) at the pressures of 5×10^{-3}–5×10^{-1} MPa.

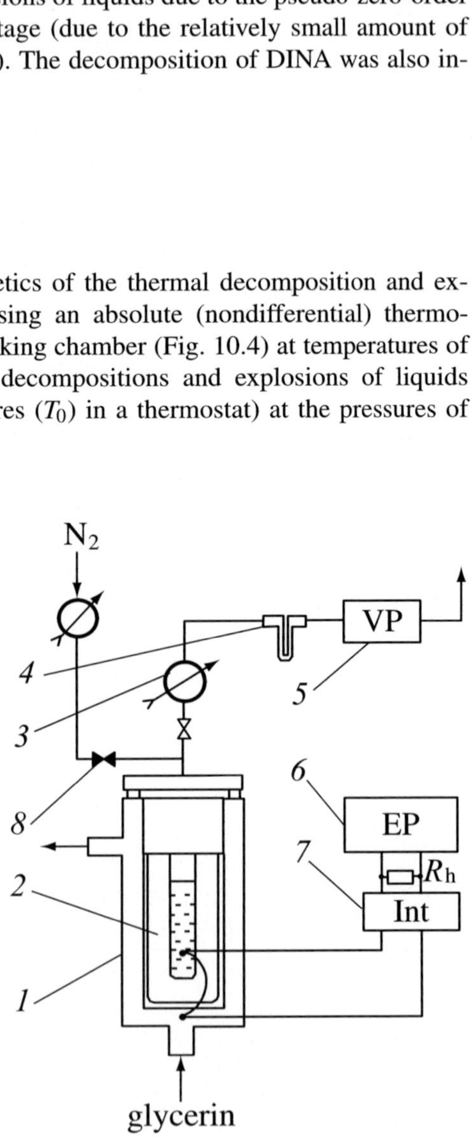

Fig. 10.4 Experimental setup: *1*, sealed flow thermostat; *2*, Dewar vessel; *3*, pressure-vacuum gauge; *4*, trap filled with liquid N_2; *5*, vacuum pump; *6*, electronic potentiometer; *7*, intensifier; *8*, valve for high-pressure line

10.5.2 Experimental Results: Kinetics of the Initial Stages of DINA Decomposition

In this case, the decomposition thermograms are characterized by a clearly defined plateau in the initial part followed by a slow monotonous dip. The thermal effect of the reaction was found by integrating the thermograms to be $Q = 3.9 \pm 0.2\,\text{kJ}\,\text{cm}^{-3}$.

The kinetics for the initial stage of the decomposition at various pressures, obtained using $Q = 3.9\,\text{kJ}\,\text{cm}^{-3}$, is illustrated in Fig. 10.5. As the pressure grows, the decomposition rate at T = const increases, while the effective activation energy decreases from $E \approx 187.5\,\text{kJ}\,\text{mol}^{-1}$ at $P_\infty = 5 \times 10^{-3}$ MPa to $E \approx 145.8\,\text{kJ}\,\text{mol}^{-1}$ at $P_\infty = 10^{-1}$ MPa. These two limiting values are in good agreement with literature data on the kinetics of DINA decomposition obtained under significantly different conditions for product removal. The first value ($187.5\,\text{kJ}\,\text{mol}^{-1}$) was obtained in [8] for decomposition in a Bourdon device characterized by very small m/v ratios (where m is the sample weight and v is the working chamber volume). The second value ($145.8\,\text{kJ}\,\text{mol}^{-1}$) was determined for $\eta = 0.1$–0.2 and pressures significantly exceeding 5×10^{-1} MPa (the crude estimate of the conditions used in [9] gives a minimum pressure of 15 MPa at $\eta = 0.1$).

The results obtained can be explained by the catalytic effect of the gaseous decomposition products on the overall process of DINA decomposition. These products are present in the liquid in the form of small bubbles. As the pressure in the thermograph's working chamber grows, the pressure in the "liquid–gas (reaction product)" also increases. This results in the diminution of the bubbles and, according to Stokes' law, decelerated bubble flotation. According to Henry's law, the concentration of dissolved gaseous products that catalyze the decomposition increases. These factors promote autocatalysis in conjunction with pressure growth in the sys-

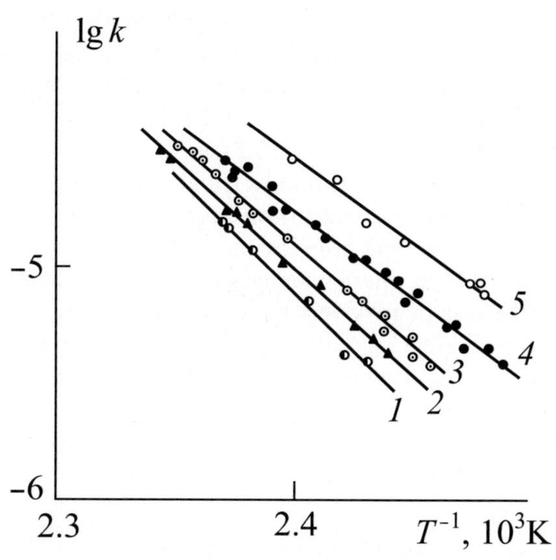

Fig. 10.5 Kinetics of the initial stages of DINA decomposition at $P_\infty = 5 \times 10^{-3}$ (*1*), 2×10^{-2} (*2*), 4×10^{-2} (*3*), 10^{-1} (*4*), 5×10^{-1} (*5*) MPa

tem. Thus one can conclude that the kinetics of the initial stage of DINA decomposition is characterized by two limiting activation energies: $E \approx 187.5\,\mathrm{kJ\,mol^{-1}}$ in the absence of autocatalysis and $E \approx 145.8\,\mathrm{kJ\,mol^{-1}}$ at maximum autocatalysis. The corresponding expressions for the rate are

$$k = 2.5 \times 10^{13} \exp(-35000/RT)[s^{-1}] \text{ at } P_\infty \geq 10^{-1}\,\mathrm{MPa}, \tag{10.14}$$

$$k = 3 \times 10^{18} \exp(-45000/RT)[s^{-1}] \text{ at } P_\infty < 5 \cdot 10^{-3}\,\mathrm{MPa}. \tag{10.15}$$

Strictly speaking, the true rate constant of the initial stage is determined by expression (10.15), since otherwise—see Eq. (10.14)—it is implied that some of the decomposition products have accumulated in the system.

10.5.3 Macrokinetics of Heat Evolution During DINA Decomposition: Thermal Explosion

The experiments used to study the critical conditions for DINA thermal explosion at various pressures were carried out using the experimental setup schematically shown in Fig. 10.4.

First, let us consider the qualitative characteristics of the process (Fig. 10.6). At low pressures ($P_\infty = 5 \times 10^{-3}$ MPa) the maximum sample warm-up value, $\Delta T_{st} = T_{max} - T_0$, does not exceed 5.3°C. As T_0 grows, ΔT_{st} initially increases, reaching its maximum value at $T_0 = 152$°C, and then it decreases, passing zero (at $T_0 =$ 179°C), so the sample temperature under stationary conditions becomes lower than the thermostat temperature T_0 ($\Delta T_{st} < 0$).

The increase in the pressure results in a dramatic change in the situation. As T_0 grows up to a certain value (T_0^{cr}), the warm-up value increases monotonously. Upon passing T_0^{cr} it increases very rapidly. For example, at $P_\infty = 2 \times 10^{-2}$ MPa and $T_0 = 145.5$°C $\Delta T_{st} = 12$ deg, and at $T_0 = 150$°C $\Delta T_{st} = 28$ deg. As T_0 is increased further, ΔT_{st} reaches its maximum value and then decreases. As the pressure increases up to $P_\infty = 4 \times 10^{-2}$ MPa, this pattern becomes more pronounced: at $T_0 = 143.5$°C $\Delta T_{st} = 11.5$ deg, and at $T_0 = 145.5$°C $\Delta T_{st} = 47.5$ deg. The lifetime of such high warm-ups is quite short (several minutes) and they reduce due to burning-out. The conversion degree corresponding to the maximum ΔT_{st} estimated using the method proposed in [10] does not exceed $\eta = 0.1$–0.2. At $T_0 > T_{0cr}$ and $P \geq 0.1$ MPa, quasi-stationary warm-up growth ceases at some point. The warm-ups in the system grow very rapidly, attaining very high values. The maximum warm-up values can be recorded with high accuracy, while the process dynamics cannot be followed by devices with response times of about 1 s (microgalvanometric amplifier and electronic potentiometer).

In this case, due to the absence of the temperature distribution over the sample volume (Bi $\ll$ 1), the classic Semenov scheme can be applied to analyze the heat evolution:

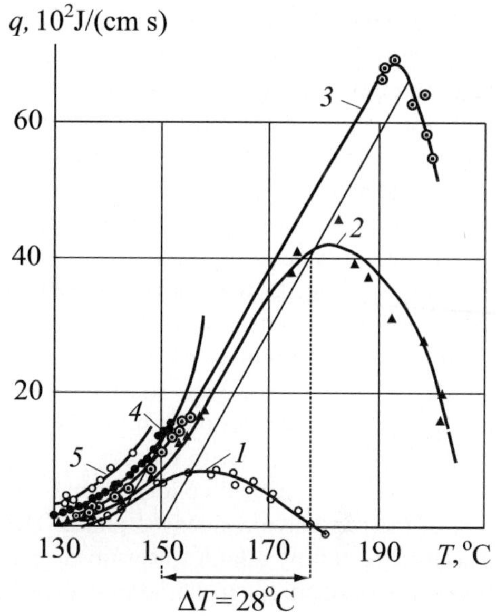

Fig. 10.6 The Semenov diagram for the thermal explosion of DINA at $P_\infty = 5\times10^{-3}$ (*1*), 2×10^{-2} (*2*), 4×10^{-2} (*3*), 10^{-1} (*4*), 5×10^{-1} (*5*) MPa

$$q(T) = \frac{\alpha S}{V}\Delta T_{\mathrm{st}}, \qquad T = T_0 + \Delta T_{\mathrm{st}}. \tag{10.16}$$

The $q(T)$ dependence (the Semenov diagram) can be plotted from the $\Delta T_{\mathrm{st}}(T_0)$ data using Eq. (10.16). The plots obtained for $P_\infty = \mathrm{const}$ are shown in Fig. 10.6.

At low T_0, the shapes of the curves are typical for the thermal explosion of non-volatile explosives (exponential temperature dependence). As T_0 increases (especially at low P_∞) the shapes of the curves change. The divergence of the curves in the low-temperature region is associated with autocatalysis with gaseous decomposition products, as discussed in the previous section (increasing the pressure increases the decomposition rate). The critical temperature at $P_\infty = 10^{-1}$ MPa and $\alpha S/V = 1.5\times10^{-2}\,\mathrm{J\,cm^{-3}\,s^{-1}\,deg^{-1}}$ is $T_0^{\mathrm{cr}} = 142.0$°C. The experimental value for the maximum warm-up in the vicinity of the explosion initiation limit is found to be $\Delta T_{\max}\,10.5$°C.

The unusual character of the $q(T)$ dependencies at low pressures and elevated temperatures can be explained using the theory of the thermal explosion of volatile explosives presented above, which takes the effect of volumetric evaporation into account. As the temperature increases, heat evolution during the decomposition of the liquid explosive occurs simultaneously with heat absorption due to liquid evaporation from the inner surface of the bubbles of the gaseous product. In this case, the heat-balance equation for stationary conditions is

$$Qk_0\exp\left(-\frac{E}{RT}\right)\left[1-\frac{L\mu_{\mathrm{v}}}{Q\mu_{\mathrm{g}}}\frac{P_0\exp(-L/RT)}{P_\infty - P_0\exp(-L/RT)}\right] = \frac{\alpha S}{V}(T-T_0). \tag{10.17}$$

At high pressures ($P_\infty \geq 10^{-1}$ MPa) and low temperatures, the term in the square brackets is close to unity, heat losses due to volumetric evaporation are negligible and the compound exhibits the behavior characteristic of a nonvolatile substance until the critical conditions are reached. As the temperature is increased (or the pressure is decreased), the effect of the volatility increases, resulting in a decrease in the total positive thermal effect. At $T_0 = T_{ad}$, determined using expressions (10.10) and (10.11), the compound overheat with respect to the medium is zero. At $T_0 > T_{ad}$ under stationary conditions the liquid is underheated ($\Delta T_{st} < 0$), since the amount of heat required for its evaporation is higher than the amount of heat released during the chemical reaction (Fig. 10.6, curve 1).

For certain combinations of the decomposition kinetic parameters and experimental conditions (see Sect. 10.3), the system performs a very rapid transition from the low warm-up mode to the high (from dozens to hundreds of degrees) warm-up mode. The boundary (with respect to the parameters) between these two modes is determined by a critical condition that depends on the kinetics of the decomposition of the explosive condensed phase. The decomposition pattern for a compound already affected by a high warm-up is mainly determined by the kinetics of vapor-phase decomposition. If the latter process is active enough and the heat and mass transfer from the system are relatively poor, "crossing the boundary in the condensed phase" will be followed by vapor self-inflammation. In the opposite case, after "crossing the boundary in the condensed phase," decomposition of the compound occurs in the "flameless flash" mode.

To conclude, let us return to the critical conditions for the thermal explosion of DINA at atmospheric pressure. Data on the critical conditions for the thermal explosion of DINA from [6] (thermostating with forced mixing) and [11] (thermostating in a Dewar vessel) treated using the Semenov expression, as well as data obtained in our work, are shown in Fig. 10.7. The straight line in Fig. 10.7 is plotted using the expression for the rate constant Eq. (10.14) obtained in our work. The good

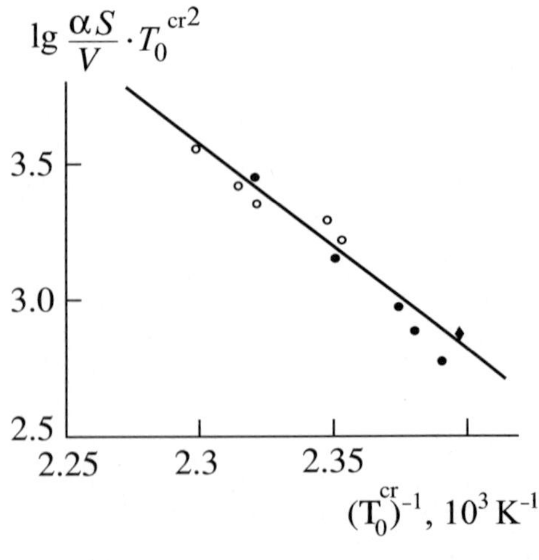

Fig. 10.7 Critical conditions for the thermal explosion of DINA at $P_\infty = 10^{-1}$ MPa based on literature data (*unfilled circles* [6], *filled circles* [11]), calculated using expression (10.14) (*straight line*), and obtained based on experimental data from our work (*filled diamond*)

agreement observed between the experimental and theoretical results confirms our suggestion of a significant autocatalytic influence from the gaseous reaction products on the thermal explosion at pressures of about 10^{-1} MPa.

References

1. Merzhanov AG, Slutsker BM, Shteinberg AS (1968) Combust Explos Shock Waves 4:312
2. Belyaev AF (1968) Combustion, detonation and work of explosion of condensed systems. Nauka, Moscow
3. Andreev KK (1966) Thermal decomposition and combustion of explosives. Nauka, Moscow
4. Merzhanov AG, Dubovitsky FI (1966) Rus Chem Rev 35:640
5. Dubovitsky FI, Manelis GB, Merzhanov AG (1958) Dokl Phys Chem 121:668
6. Dubovitsky FI, Barzykin VV, Merzhanov AG (1960) Izv Akad Nauk SSSR Ser Khim 6:361
7. Barzykin VV, Merzhanov AG (1964) Zh Fiz Khim 38:2640
8. Dubovitksy FI, Rubtsov YuI, Barzykin VV, Manelis GB (1960) Izv Akad Nauk SSSR Ser Khim 6:1126
9. Samoylenko NG, Vinokurov AA, Abramov VG, Merzhanov AG (1970) Rus J Phys Chem 44:22
10. Merzhanov AG, Abramov VG, Abramova LT (1967) Rus J Phys Chem 41:90
11. Barzykin VV, Merzhanov AG (1964) Rus J Phys Chem 38:1431

Chapter 11
High-Temperature Decomposition and Thermal Explosion of Liquid Propellant Components: Hydrogen Peroxide and Hydrazine

Abstract Highly concentrated H_2O_2 and hydrazine are energetic materials that have been widely used in rocket and space engineering. Since these liquids undergo heating in various systems over a very wide temperature range, the study of their decomposition kinetics at these temperatures is of significant importance. Original instruments for measuring very high and very low decomposition rates (corresponding to almost the entire temperature interval of practical interest) of these liquids are described. Data on the thermal explosion of highly concentrated H_2O_2 and kinetic constants for its homogeneous decomposition (corresponding to a range of decomposition rates covering six orders of magnitude) are given for the first time. Data on the rate of heterogeneous decomposition of highly concentrated H_2O_2 on surfaces of various practically interesting construction materials are presented. Similar data for hydrazine are also given in this chapter.

11.1 Introduction

This chapter is devoted to a discussion of the macrokinetics and kinetics of the thermal decomposition of highly concentrated liquid hydrogen peroxide (h/c H_2O_2) and N_2H_4 (hydrazine), which were selected from numerous liquid propellants for reasons given below. Due to its high-energy and performance characteristics, h/c H_2O_2 has been used as an oxidizer in rocket engines and torpedoes as well as a monopropellant in the catalytic units of space vehicle orientation and touchdown engine systems for quite a long time. In the past few years its range of applications has been significantly widening since, from an environmental safety point of view, h/c H_2O_2 has and will continue to have few competitors. Since the only products of its decomposition are oxygen and water, h/c H_2O_2 has been used in the life-support systems of spacecraft. From a practical point of view it is worth noting that the high reactivity of h/c H_2O_2means that if it is accidentally spilled it quickly converts into water. However, the use of h/c H_2O_2 is also associated with some problems due to its low thermal stability and ability to decompose catalytically upon contact with quite inert metals. Based on his personal experience of working in committees that have investigated many accidents, the author of this book came to the conclusion that h/c H_2O_2 can be considered the most dangerous (with respect to explosions and fire hazards) liquid propellant component, especially in monopropellants. This conclusion is backed up by the following relatively recent serious accidents.

An accident in the Russian submarine *Kursk* which sank in the Barents Sea involved an h/c H_2O_2-fueled torpedo explosion. Before this accident, on 18 March 1980, an A-7 rocket exploded during fueling, killing 48 people at the Plesetsk spaceport [1]. As was discovered by the author of this book, the accident was caused by the thermal explosion of h/c H_2O_2 which came into contact with catalytically active Pb-containing solder. The soldering alloy was mistakenly used by a manufacturer to solder a stainless steel screen filter in the h/c H_2O_2 supply line. The author was honored with a Plesetsk Spaceport Diploma for investigating the accident and developing a number of practically important recommendations regarding fire and explosion safety when using h/c H_2O_2. Nowadays, since the widespread use of h/c H_2O_2 has meant that it must be stored for a long time in tanks of various machines and devices, the issue of its heterogeneous decomposition upon contact with surfaces of construction materials is still very important. Data on this matter is scarce in the literature.

In contrast to h/c H_2O_2, at long-term storage temperatures liquid hydrazine exhibits high thermal stability. Because of this, as well as its high-energy characteristics, hydrazine is widely used for various applications (for example, in the Space Shuttles) and is considered a promising material for use in future space engineering.

However, in order to be able to analyze the performance of hydrazine in units that are used for several years, data on its low-temperature homogeneous and heterogeneous decompositions upon contact with construction materials of tanks, pipelines, etc., are required. In addition, it is also important to obtain information on the high-temperature decomposition and thermal explosion of hydrazine. This necessity is explained by considering the characteristics of a catalytic unit operating according to a standard cyclogram. During a stop phase, monopropellant is not supplied to the unit. Under these conditions, hydrazine in the cutoff valve and in the supply pipeline rapidly heats up due to the heat from the cooling (but still burning hot) catalytic unit. Without employing special measures in order to suppress the heating-up effect, thermal explosions of hydrazine can occur.

Such a scenario was discussed in [2], where the increased explosion hazard presented by hydrazine use in the Space Shuttle was emphasized (hydrazine is used in a turbo-compressor unit that drives the hydraulic system).

11.2 Some Additional Experimental Techniques

The kinetics of the thermal decomposition and explosion of h/c H_2O_2 and hydrazine was studied using a method described in Sect. 10.5, a manometric method developed by Prof. Nechiporenko (Institute of Chemical Physics Problems of the Russian Academy of Sciences), and techniques using specially designed kinetic devices (automatic volumetric device (AVD) and Microdroplet).

The use of the AVD and Microdroplet devices allows one to obtain information about the decomposition macrokinetics over a very wide temperature range: high temperatures correspond to conditions in the pipelines of a liquid rocket engine cooling system, while moderate and relatively low temperatures are characteristic of monopropellant storage. In contrast to the kinetic characteristics of dilute aqueous solutions of H_2O_2, data on the homogeneous and heterogeneous decompositions of h/c H_2O_2 are rare and inconsistent. The corresponding data for hydrazine are only reported in [3] and [4].

11.2.1 Automatic Volumetric Device (AVD)

The main part of the device is an automated gas burette with a compensation vessel that moves via a reversible motor. A U-tube gauge responding to the pressure change in the sealed system in which decomposition of the liquid takes place is

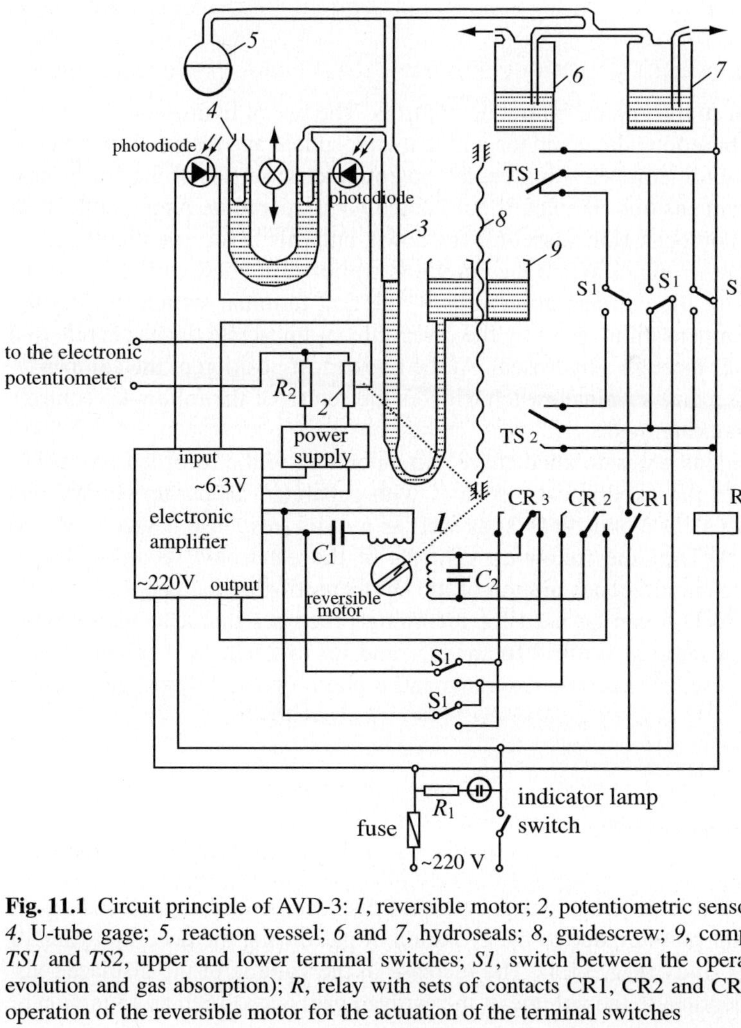

Fig. 11.1 Circuit principle of AVD-3: *1*, reversible motor; *2*, potentiometric sensor; *3*, gas burette; *4*, U-tube gage; *5*, reaction vessel; *6* and *7*, hydroseals; *8*, guidescrew; *9*, compensation vessel; *TS1* and *TS2*, upper and lower terminal switches; *S1*, switch between the operation modes (gas evolution and gas absorption); *R*, relay with sets of contacts CR1, CR2 and CR3 controlling the operation of the reversible motor for the actuation of the terminal switches

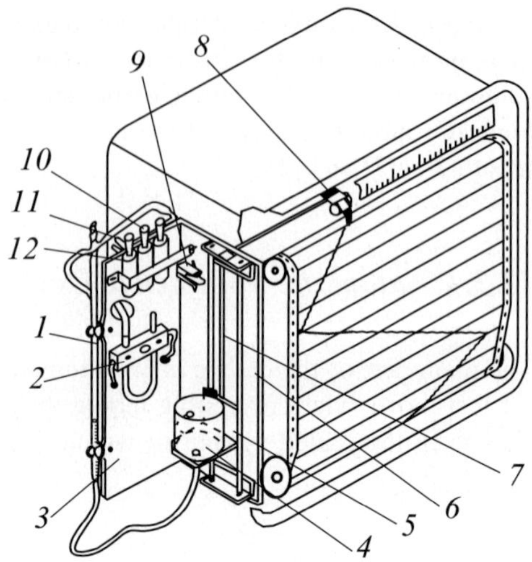

Fig. 11.2 Circuit principle of AVD-5: *1*, gas burette; *2*, manometer and photocell; *3*, baseplate; *4*, compensation vessel; *5*, compensation vessel tray; *6*, guiding rod; *7*, thread connecting the compensation vessel tray with the electronic recorder slide; *8*, recorder slide; *9*, terminal switch; *10–12*, hydroseals

used as a sensor in the engine follow-up controls. The use of hydroseals in the line connecting the burette volume and the ambient atmosphere and terminal switches in the reversible motor–compensation vessel–potentiometric sensor of the compensation vessel system ensures completely automated AVD operation. As gas collects in the system the compensation vessel moves down, and this is recorded on the electronic potentiometer chart. When the burette is completely filled with the gas (the lowest position of the compensation vessel), the first terminal switch triggers the lifting of the compensation vessel by the reversible motor, while the gas is released to the ambient air through a hydroseal. At the uppermost position of the compensation vessel the second terminal switch triggers the return of the follow-up controls to the normal working mode.

Thus one obtains a saw-toothed curve corresponding to the recorded process of gas evolution. In the experiments, devices with separated follow-up control and recording units (AVD-2 and AVD-3) as well as a quick-response device (AVD-5) were used. In AVD-5, the follow-up controls of the electronic recorder are responsible for the simultaneous motion of the compensation vessel and the recorder printing head. AVD-5 can be used for recording processes that are characterized by gas-evolution rates of up to $40\,\mathrm{ncm}^3\,\mathrm{s}^{-1}$, and up to $2\,\mathrm{ncm}^3\,\mathrm{s}^{-1}$ when AVD-2 and AVD-3 are used. Schematic diagrams and a photo of the devices are given in Figs. 11.1–11.3.

11.2.2 Microdroplet Device

The operation of this device is based on liquid displacement from a closed vessel kept at a constant pressure due to the evolution of gas during the decomposition of the liquid under study (Fig. 11.4). The increase in the volume of the displaced liquid (which is identical to the volume of the evolved gas) is recorded as a function of

Fig. 11.3 Photo of AVD-3

time. The liquid displaced from the vessel connected to the reaction flask drops from a capillary to a collector. The amount of gas in the reaction flask is determined by periodically weighing the collector. For a wide range of low rates of liquid displacement the duration of droplet growth on the capillary top significantly exceeds the duration of the nonstationary formation of the droplet bottleneck in its upper part. At the moment of droplet detachment, its weight (and volume) does not depend on the displacement rate and it is completely determined by the capillary parameters and properties of the liquid (surface tension and density). This circumstance allows one to use this device to measure extremely low rates of gas evolution by recording the detachment times of individual droplets. A single droplet volume can be determined with high accuracy by periodically weighing the collector, which contains a known number of droplets.

In these experiments, a photocell installed near the capillary top was used. A simple integrator was used to record the number of impulses as a function of time.

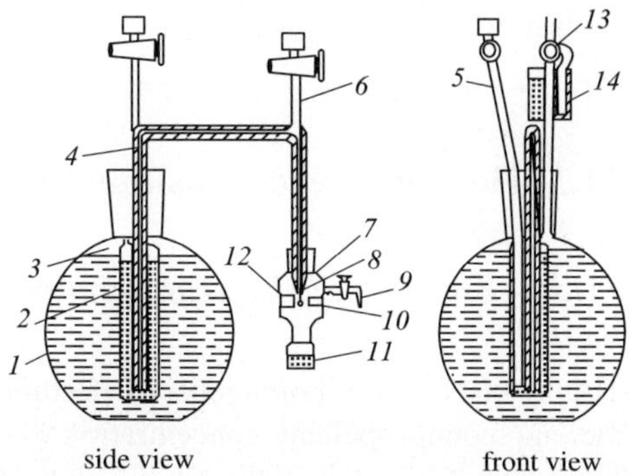

Fig. 11.4 Schematic diagram of the Microdroplet device: *1*, reaction flask; *2*, vessel with liquid being displaced; *3*, gas; *4*, outlet tube; *5*, tube for filling vessel 2; *6*, tube for liquid supply during calibration of the device; *7*, registration unit; *8*, capillary; *9*, stopcock for pressure equalization; *10*, photodiode pocket; *11*, displaced liquid collector; *12*, lighting pocket; *13*, hydroseal; *14*, hydroseal vessel

Thus gas evolution in the Microdroplet device was recorded as a step-like curve with steps of the equal height and varying width.

A high device sensitivity is ensured by applying perfluorinated liquids characterized by a low surface tension and forming very small droplets of a constant volume ($V \leq 10^{-3}\,cm^3$). It is easy to show that at a displacement rate of one droplet per hour and a decomposing liquid volume of $1000\,cm^3$ (the amount of gas evolved per volume unit for h/c H_2O_2 is almost identical to that of hydrazine), decomposition occurring with the rate constant of $k \approx 5 \times 10^{-13}\,s^{-1}$ is recorded. Such a rate constant is approximately 50 times lower than that of the homogeneous decomposition of the most stable h/c H_2O_2 samples under real conditions at a temperature of 20°C. Express methods for monitoring h/c H_2O_2 and hydrazine stability under conditions of long-term storage were developed based on the described approach.

The kinetics of the thermal decomposition and explosion of hydrazine were also studied by a manometric method mentioned above and by a thermograph-based setup similar to that described in Sect. 10.5.1 with an electrically heated thermostat and sealed glass ampoules containing hydrazine.

Before discussing the main results of the study of the thermal decomposition of h/c H_2O_2 and hydrazine, some comments about the reaction mechanisms should be made. Data on the decomposition mechanism of highly concentrated liquid hydrazine are rare in the literature, although the kinetics of the vapor were thoroughly investigated by Eberstein and Glassman [5] and Manelis and coworkers [6, 7]. The h/c H_2O_2 decomposition mechanism was studied more actively, but most works have been devoted to the low-temperature process [8, 9, 10, 11]. The most important contribution to studying the mechanism of h/c H_2O_2low-temperature decomposition was made by Purmal and coworkers [12, 13]. Some of our experimental and theoretical research was also devoted to gaining a better understanding of the chain mechanism of h/c H_2O_2 decomposition; in particular, to the qualitative analysis of the role of iron ions in the process [14]. However, since this book is mostly devoted to macrokinetic aspects of decomposition and explosion processes, material on the mechanisms of h/c H_2O_2 and hydrazine decomposition is not presented here.

Thus the topics considered in this chapter are as follows: determination of the formal kinetic parameters of the liquid-phase decompositions of the monopropellants over the widest possible temperature range; analysis of the effects of practically important construction materials on the process kinetics; an experimental study of the thermal explosion initiation mechanism and comparison of the experimental and theoretical data.

11.3 Thermal Decomposition of h/c H_2O_2

11.3.1 Kinetics of the Homogeneous Decomposition of h/c H_2O_2

The kinetics of the homogeneous decomposition of h/c H_2O_2 (samples of commercial monopropellant, concentration 94–97%) were studied over the temperature range 25–180°C. Volumetric methods (AVD, Microdroplet and others) were used at 25–90°C and atmospheric pressure, while the calorimetric (thermographic) method (see Fig. 10.4) was employed at temperatures exceeding 100°C and elevated pressures (1.5–4.0 MPa) according to the "nonvolatility" requirement (see Eq. 10.8a).

To identify the effect of the reaction vessel material on the h/c H_2O_2 decomposition, a number of experiments involving several materials at various F/V ratios were carried out (F is the area of the vessel's inner surface, V is its volume). For vessels made of molybdenum glass, opaque quartz and (to a lesser degree) pyrex glass, the F/V ratio was found to significantly influence the h/c H_2O_2 decomposition rate. In contrast, in the cases of Teflon and optical grade quartz, the decomposition rate was close to its minimum value, and a 3–4-fold change in the F/V ratio did not affect it. Thus, under these conditions the liquid phase decomposition of h/c H_2O_2 occurs as a homogeneous process, while the contribution of heterogeneous decomposition on the vessel walls is negligible (Fig. 11.8).

The thermograms obtained were analyzed using the method developed in [15].

A typical thermogram obtained in the experiments aimed at determining the thermal effect of the reaction, Q, is given in Fig. 11.5. The experimental value of $Q = 2.75 \pm 0.13\,\mathrm{kJ\,g^{-1}}$ was found to be close to that calculated from thermochemical data ($Q = 2.88\,\mathrm{kJ\,g^{-1}}$).

The initial and final H_2O_2 concentrations were determined with a minimum accuracy of 0.3% from the refraction index data obtained using a precision reflectometer IPF-23 (by using tables [8]).

The decomposition of h/c H_2O_2 was found (from thermogram data) to be a first-order reaction ($n = 0.9$–1.0). Data corresponding to a temperature of 140°C are presented in Fig. 11.6 as an example.

Due to the fact that the isothermal decomposition reaction without self-acceleration exhibited rather long periods of quasi-stationary character (see for example Fig. 11.7), most of the thermographic experiments were performed according to the following simple scheme. The liquid with initial concentration c_0 was kept at a constant temperature T_0 up to the point that the quasi-stationary mode, characterized by a constant (maximum) warm-up ΔT_{st}, began. Then the experiment was terminated, the liquid was cooled and the final concentration c_{fin} was measured. The final conversion degree, η_{fin}, was calculated from the value of c_{fin}. The correctness

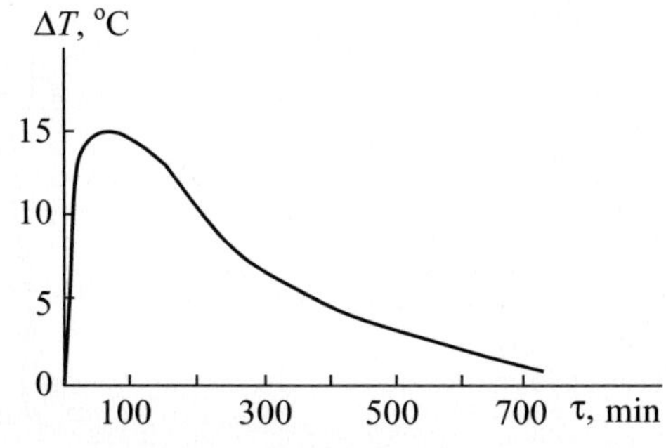

Fig. 11.5 Typical thermogram obtained for decomposition of h/c H_2O_2 at $P_\infty = 1.6\,\mathrm{MPa}$, $T_0 = 149.5$°C, $\eta_{fin} = 0.9$, $\alpha S/V\rho = 8.3 \times 10^{-3}\,\mathrm{J\,g^{-1}\,s^{-l}\,deg^{-1}}$

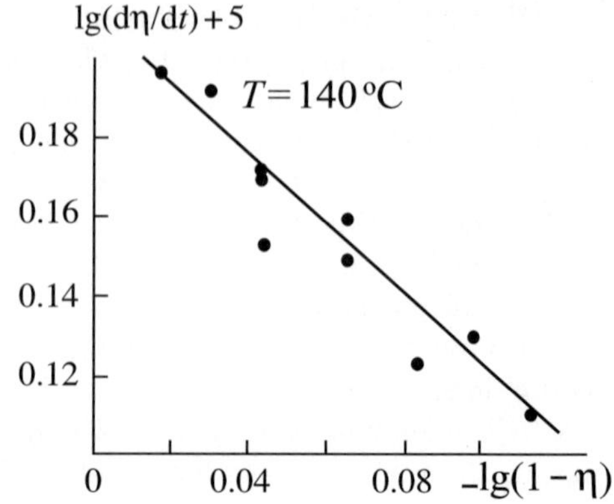

Fig. 11.6 Determining the reaction order for the thermal decomposition of h/c H_2O_2 at $T = 140°C$

of η_{fin} was double-checked by integrating the thermogram. For short experiments (40–50 min), the conversion degree is very low and its value should only be taken as accurate when fast experiments are performed at high temperatures. h/c H_2O_2 decomposition constants covering 1.5–2 orders of magnitude were calculated as $k = \dfrac{\alpha S \Delta T_{st}}{QV(1-\eta_{fin})}$ from thermographic data obtained using photovoltaic amplifiers in combination with a precision ultrathermostat from Brabender. The kinetic parameters estimated from the data plotted in Fig. 11.8 are $E = 85.8\,\text{kJ}\,\text{mol}^{-1}$ and $k_0 = 4.5 \times 10^5\,\text{s}^{-1}$.

The studied h/c H_2O_2 samples were characterized by different values of the initial concentration and the so-called thermal stability index (TSI, equal to the amount of O_2 expressed in ncm^3 evolved from 50 ml of H_2O_2 in a glass vessel at $T = 96°C$ for 1 h [16]). Since the TSI is currently widely used to qualitatively evaluate the h/c H_2O_2 stability, the analysis of the relationship between the TSI and kinetic constants of homogeneous decomposition was of significant interest. The results of this analysis are shown in Figs. 11.9 and 11.10. The TSI did not influence the

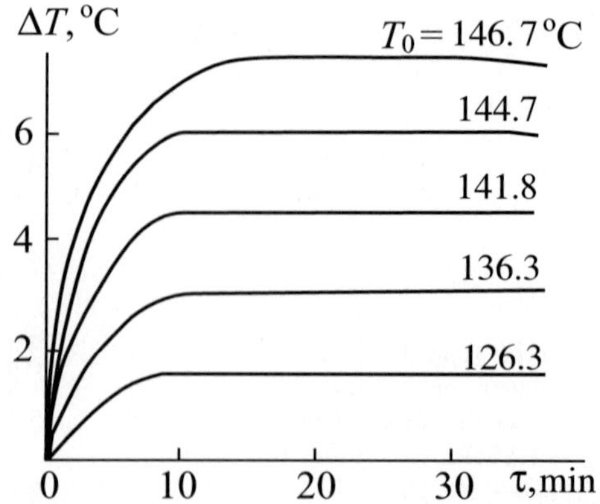

Fig. 11.7 Initial parts of thermograms for h/c H_2O_2 decomposition at various temperatures and $P_\infty = 1.6\,\text{MPa}$

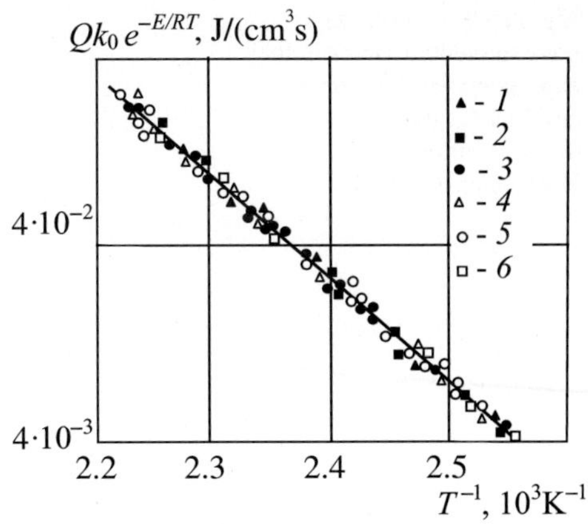

Fig. 11.8 Arrhenius plot of the rate of heat evolution during the decomposition of h/c H_2O_2 (96%, TSI = 14) in optical grade quartz ($F/V = 1.3$ (*1*), 1.9 (*2*), 4.0 (*3*) cm^{-1}) and Teflon ($F/V = 1.3$ (*4*), 4.0 (*5*), 2.0 (*6*) cm^{-1}) vessels at $P_\infty = 1.6$ MPa. The data were obtained by the thermographic method

effective activation energy. In contrast, the preexponential factor in the expression for the rate constant linearly increases as the TSI increases.

Thus one reaches the conclusion that the TSI reflects the contribution from the homogeneous process to the decomposition rate. Data on decomposition kinetics obtained for h/c H_2O_2 samples with various initial TSI values can be plotted in Arrhenius coordinates as one straight line with a normalized TSI.

The final Arrhenius plot generalizing the experimental data on high- and low-temperature decomposition of h/c H_2O_2 obtained using thermographic and volumetric methods is shown in Fig. 11.11 (the data are normalized to TSI = 14). The experimental points, which cover a wide temperature interval (25–180°C) and a wide range of homogeneous decomposition rates ($\sim$ 6 orders of magnitude), fall on

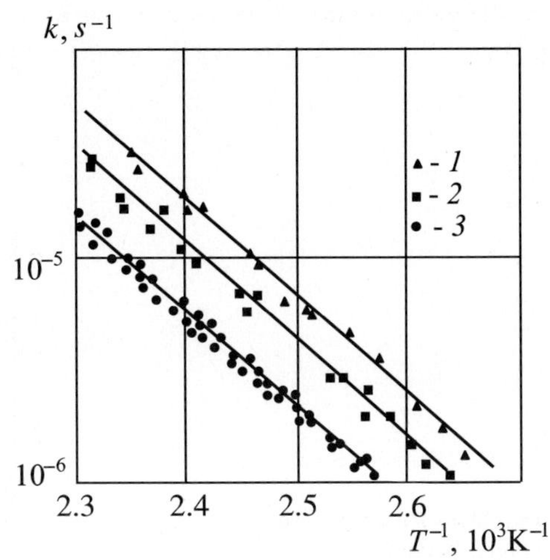

Fig. 11.9 Decomposition kinetics for h/c H_2O_2 with TSI = 26–30 (*1*), 17–19 (2) and 9–11 (*3*)

Fig. 11.10 Plot of the pre-exponential factor in the expression for the rate constant for the decomposition of h/c H_2O_2against its TSI

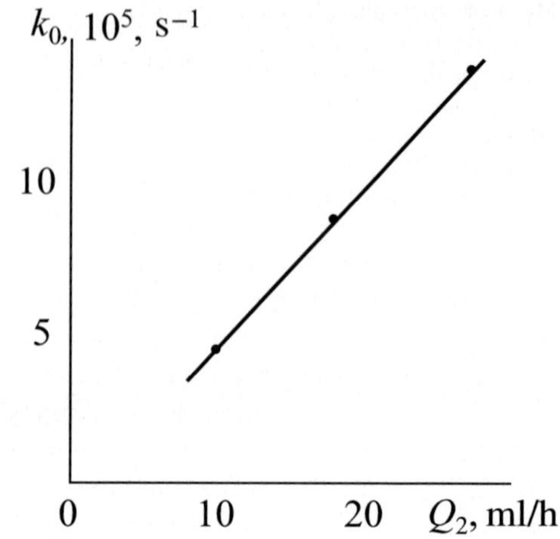

straight line. This indicates that the kinetic mechanism is the same at all of these temperatures.

The expression for the rate constant of the homogeneous decomposition of h/c H_2O_2 is

$$k = 4.8 \times 10^5 \exp\left(-\frac{20800}{RT}\right) [s^{-1}]. \quad (11.1)$$

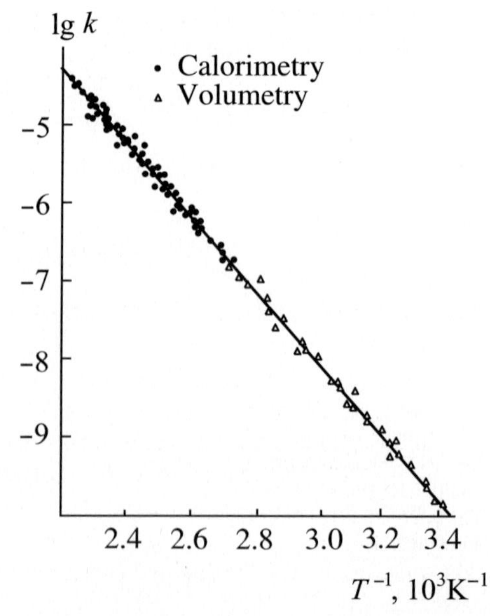

Fig. 11.11 Kinetics of the homogeneous thermal decomposition of h/c H_2O_2 (94–98%), TSI = 14

11.3.2 Kinetics of the Heterogeneous Decomposition of h/c H_2O_2 on the Surfaces of Construction Materials

The decomposition of H_2O_2 on the surfaces of catalytically active materials has been very actively studied, while data on the decomposition of h/c H_2O_2 in the presence of materials characterized by low catalytic activity are rare. There are no data on the kinetic parameters for decomposition at temperatures exceeding 100°C. This information is very important (especially for high F/V ratios) for analyzing the behavior of h/c H_2O_2 during long-term storage in metal containers and when it is used as a heat carrier in the cooling lines of liquid rocket engines.

To a first approximation, the decomposition of H_2O_2 in the presence of metal foil can be considered to be a combination of homogeneous and heterogeneous processes. The macrokinetics of the heterogeneous process depends on the nature of the metal and its active surface area. One can estimate the kinetic constants for the heterogeneous reaction on the metal surface by measuring the total decomposition rate and subtracting the component corresponding to the homogeneous reaction.

Samples of 96–98% H_2O_2 were studied in the experiments. All of the characteristics were measured after the system had attained a stationary mode characterized by constant rates of heat and gas evolution. Due to relatively low decomposition rates (especially in volumetric experiments at low temperatures), the H_2O_2 concentration hardly changed during the experiment. The heterogeneous decomposition of h/c H_2O_2 at low temperatures was studied by a volumetric method using an AVD-type device connected to a thermostated reaction vessel (25 cm^3, optical grade quartz) installed on a vibration stirrer. The heat carrier was pumped from an ultrathermostat and circulated in the reaction vessel water jacket.

Two rates of gas evolution were measured: before a small piece of metal foil ($F = 10$–$50\,cm^2$) was placed into the reaction vessel with h/c H_2O_2 (W_{hom}) and after this (W_{Σ}). Assuming that

$$W_{\Sigma} = k_{\Sigma}V = k_{hom}V + k_{S}F, \tag{11.2}$$

the rate constant for heterogeneous decomposition was calculated as

$$k_{S} = \frac{W_{\Sigma} - W_{hom}}{11200(G/\mu)(F/V)}\,[\mathrm{cm\,s^{-1}}], \tag{11.3}$$

where G is the h/c H_2O_2 weight and $\mu_{H_2O_2}$ is its molecular weight.

The high-temperature experiments were carried out using the calorimetric device used to study the kinetics of homogeneous decomposition described above. A small piece of the metal under study was placed in the reaction vessel.

The first set of experiments was performed using samples of stainless steel foil (grade H18N10T, 0.15 mm thick, $F = 6.3$–$28\,cm^2$) and aluminum foil (grade AD-1, 0.1 mm thick, $F = 24\,cm^2$). Generally, when all other conditions were the same, the presence of the metal was found to result in a significant increase in the rate of h/c H_2O_2 decomposition for the indicated range of metal surface areas.

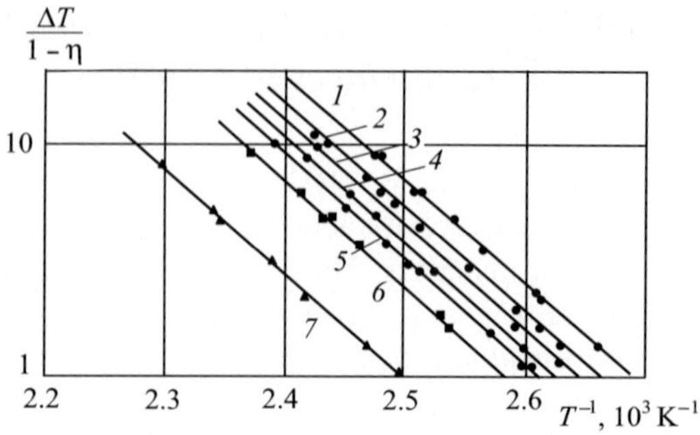

Fig. 11.12 Linearization of data on the stationary warm-up during h/c H_2O_2 decomposition in the presence of stainless steel (H18N10T) at $F/V = 6.2$ (*1*), 4.2 (*2*), 2.9 (*3*), 2.1 (*4*), 1.3 (*5*) cm^{-1}, aluminum-based alloy (AD-1) at $F/V = 5\,cm^{-1}$ (*6*) and in the absence of any metal (*7*)

Data on the dependence of the stationary warm-up on the liquid temperature for the homogeneous decomposition of H_2O_2 (7) and for the processes occurring in the presence of stainless steel foil (1–5) and aluminum AD-1 (6) are given in Fig. 11.12.

Noting that heterogeneous decomposition takes place in the kinetic region (see below) according to a first-order reaction occurring in the quasi-stationary mode ($c\rho dT/dt \ll \alpha S \Delta T/V$), one can write the heat-balance equation as

$$Qk_0 \exp\left(-\frac{E}{RT}\right)(1-\eta) + Qk_{0S} \exp\left(-\frac{E_S}{RT}\right)(1-\eta)\frac{F}{V} = \frac{\alpha S}{V}\Delta T_\Sigma. \qquad (11.4)$$

If the kinetics of the homogeneous decomposition of h/c H_2O_2 in the presence of a metal is identical to that of the "pure" h/c H_2O_2 (7), then one can calculate the constant for the heterogeneous decomposition rate using the following modified form of Eq. (11.4):

$$k_S = \frac{\Delta T_\Sigma - \Delta T_{hom}}{Q(F/V)(1-\eta)} \frac{\alpha S}{V}. \qquad (11.5)$$

The corresponding analysis of the results obtained is illustrated by Figs. 11.13 and 11.14.

Studying the effect of trace iron ion on the rate of homogeneous h/c H_2O_2 decomposition is an individual research task. A plot of the total rate constant for the

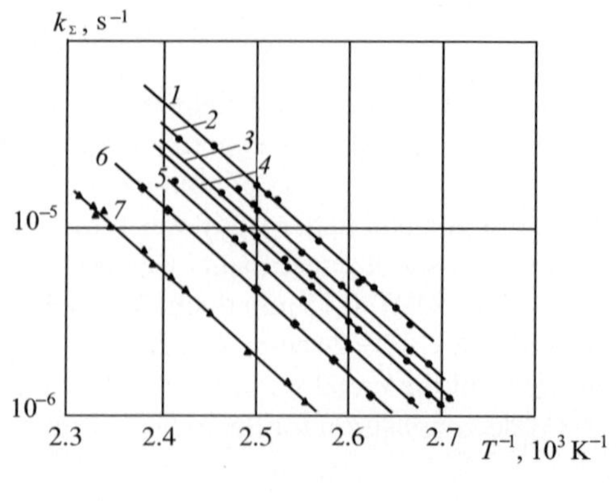

Fig. 11.13 Plot of the total rate of h/c H_2O_2 decomposition in the presence of stainless steel (H18N10T) at $F/V = 6.2$ (*1*), 4.2 (*2*), 2.9 (*3*), 2.1 (*4*), 1.3 (*5*) cm^{-1}, aluminum-based alloy (AD-1) at $F/V = 5\,cm^{-1}$ (*6*) and in the absence of any metal (*7*)

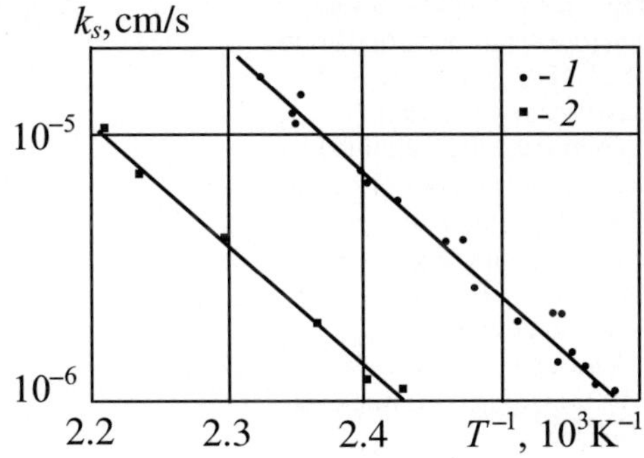

Fig. 11.14 Kinetics of the heterogeneous decomposition of h/c H_2O_2 on stainless steel H18N10T (*1*) and aluminum-based alloy AD-1 (*2*) surfaces

thermal decomposition of h/c H_2O_2 against the stainless steel foil surface area is given in Fig. 11.15. The value of the constant obtained by extrapolation to $F = 0$ slightly exceeds that of the pure h/c H_2O_2. However, the inaccuracy of the calculated constants of heterogeneous decomposition resulting from this effect is quite small at $k_\Sigma F/V \gg k_{hom}$.

The described approach was used to study the heterogeneous decomposition of h/c H_2O_2 on aluminum (grades AD-1 and A-000-93), niobium and nickel-based alloy (grade NP-2 before and after oxidation) surfaces.

The experimental results processed using Eqs. (11.3) and (11.5) in the Arrhenius coordinates are presented in Fig. 11.16. The corresponding expressions for the constants of heterogeneous decomposition k_{0S} and E_S are summarized in Table 11.1.

The volumetric experiments were performed at 30–90°C, and the calorimetric experiments were performed at 80–160°C.

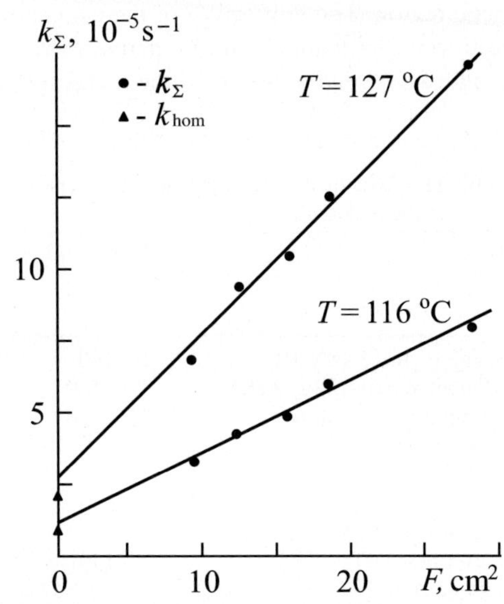

Fig. 11.15 Plots of the total rate of decomposition of h/c H_2O_2 against the stainless steel (H18N10T) surface area

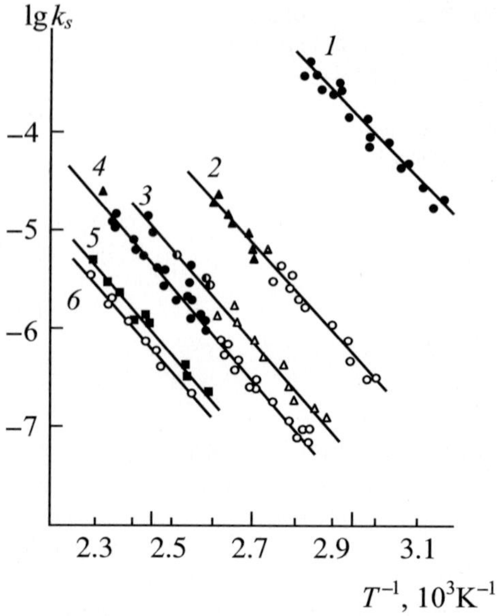

Fig. 11.16 Kinetics of the heterogeneous decomposition of h/c H_2O_2 on surfaces of construction materials: niobium, the data obtained by calorimetry (*1*); nickel-based alloy NP-2 (unoxidized) (*2*), the data obtained by calorimetry (*filled triangles*) and volumetry (*unfilled circles*); nickel-based alloy NP-2 (oxidized) (*3*), the data obtained by calorimetry (*filled circles*) and volumetry (*unfilled triangles*); stainless steel H8N0T (*4*), the data obtained by calorimetry (*filled circles*) and volumetry (*unfilled circles*); aluminum-based alloys AD-1 (*5*) and A-000-93 (*6*), the data obtained by volumetry

The good agreement observed between the experimental data obtained by the two methods, in particular for the stainless steel and the oxidized and unoxidized nickel, implies that the macrokinetic mechanisms of decomposition on the surfaces of these metals were the same over a wide temperature range. The decomposition rate was found to be highest for niobium and nickel (Fig. 11.16) and was independent of the stirring rate (Fig. 11.17).

Thus one can conclude that the heterogeneous decomposition occurs in the kinetic region. The uniformity of the concentration is ensured by molecular diffusion at lower temperatures and by active convection caused by the growth and detachment of oxygen bubbles near the metal surface at higher temperatures.

Table 11.1 Kinetic constants for the heterogeneous decomposition of h/c H_2O_2 on the surfaces of construction materials

Material	Temperatures, °C	Preexponential factor k_{0S}, cm s^{-1}	Activation energy E_S, kJ mol^{-1}
Stainless steel H18N10T	60–160	5×10^5	86.7
Aluminum-based alloy AD-1	110–160	1.3×10^5	87.5
Aluminum-based alloy A-000-93	110–160	1.0×10^5	87.5
Niobium	40–90	9×10^8	83.3
Nickel-based alloy NP-2 (non-oxidized)	50–110	1.5×10^7	86.7
Nickel-based alloy NP-2 (oxidized)	70–140	2.0×10^6	86.7

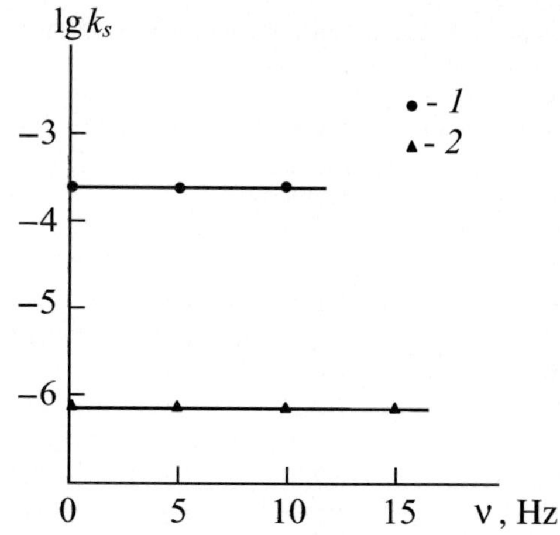

Fig. 11.17 Effect of the stirring rate on the rate of h/c H_2O_2 heterogeneous decomposition on the surfaces of niobium (*1*) and nickel-based alloy NP-2 (*2*) at $T = 72°C$

The effective decomposition activation energies of these very different (in terms of their physicochemical properties) metals are in excellent agreement (within the experimental accuracy $E_S = 85.4 \pm 2.1\,\text{kJ mol}^{-1}$). This value is almost equal to the activation energy of the homogeneous decomposition of pure h/c H_2O_2 ($E = 86.7\,\text{kJ mol}^{-1}$). The significant difference between the preexponential factors (see Table 11.1), which reaches, for example, four orders of magnitude in the case of aluminum and niobium, seems to be due to the difference in the numbers of reaction chains that form on the surfaces of these metals.

From a macrokinetics viewpoint, the mechanism of the heterogeneous decomposition of h/c H_2O_2 on the surfaces of the studied metals can be described as follows. There are a certain number of initiation centers for the unbranched chain reaction of H_2O_2 decomposition. Their surface concentration depends on the physicochemical nature of the metal and probably on the surface pretreatment method. One can assume the validity of relationship $k_{0S} = AC_S$, where C_S [cm^{-2}] is the surface concentration of initiation centers, which depends on the metal, and A [$\text{cm}^3\,\text{s}^{-1}$], the constant related to the homogeneous decomposition of H_2O_2. After the process initiates on the metal surface (the free radicals $HO_2\cdot$ and $HO\cdot$ appear), the chain reaction develops into the liquid phase. Since the homogeneous decomposition of H_2O_2 is characterized by relatively long reaction chains, the subsequent decomposition occurs according to a homogeneous mechanism [14]. This explains the equality between the heterogeneous E_S/E and homogeneous constants.

Despite the high rate of the process in the vicinity of the metal surface in comparison with that in the volume of the liquid phase, the corresponding H_2O_2 concentrations are similar due to a very low thickness of the interface reaction layer and thus relatively high mass-transfer rates. This seems to explain the independence of the k_S value of the stirring rate.

Based on the suggested mechanism, one can suppose that stabilizing agents for the homogeneous decomposition of H_2O_2 are also efficient for the

Table 11.2 Effect of phenol stabilizer (10^{-2}% PhOH) on the kinetics of the heterogeneous decomposition of h/c H_2O_2

Temperature, °C	Material	F/V, cm^{-1}	k_S, $cm\,s^{-1}$	k_S(PhOH), $cm\,s^{-1}$	$\frac{k_S}{k_S(\text{PhOH})}$
90.5	H18N10T	1.25	1.8×10^{-7}	8.05×10^{-8}	2.24
81.1	H18N10T	1.25	9.5×10^{-8}	4.32×10^{-8}	2.20
96.2	H18N10T	1.25	2.5×10^{-7}	1.11×10^{-7}	2.25
83.1	H18N10T	1.25	8.3×10^{-8}	3.45×10^{-8}	2.41
98.4	H18N10T	1.25	2.5×10^{-7}	1.06×10^{-7}	2.34
67	NP-2	2.5	4.6×10^{-7}	1.88×10^{-7}	2.44
62.5	NP-2	2.5	3.1×10^{-7}	1.34×10^{-7}	2.31
79.8	NP-2	2.5	1.7×10^{-6}	7.3×10^{-7}	2.39
85.3	NP-2	2.5	2.5×10^{-6}	1.10×10^{-6}	2.26
88	NP-2	4.1	4.1×10^{-6}	1.78×10^{-6}	2.30

pseudo-heterogeneous processes under consideration. The correctness of this suggestion has been shown experimentally, at least for stainless steel and nickel-based alloy, by using phenol-stabilized h/c H_2O_2 (see Table 11.2). The introduction of phenol (10^{-2}%) results in a 2.2–2.5-fold decrease in the rate of h/c H_2O_2 homogeneous decomposition. The decreases in the rate constants of heterogeneous decomposition on the stainless steel and nickel-based alloy surfaces are identical within experimental accuracy.

11.3.3 Thermal Explosion of H_2O_2

The thermal explosion kinetics and corresponding critical conditions were studied by the method described in Sect. 10.5.3. In the experiments carried out at $P_\infty =$ 0.1–1.6 MPa, samples of h/c H_2O_2 characterized by initial TSI values of 14–30 were studied.

Since typical thermograms obtained at high pressures (Fig. 11.7) have parts attributed to the stationary process, the corresponding points can be plotted on a Semenov diagram (Fig. 11.18). The data obtained at various P_∞ values are in qualitative agreement with the $q(T)$ dependencies given in Figs. 10.2 (theory) and 10.6 (heat evolution during the decomposition of DINA). The congruence of the low-temperature parts of the heat-evolution curves for low and high pressures is in quantitative agreement with the theory.

The congruence of the low-temperature parts at $P_\infty =$ const for h/c H_2O_2 decomposition is explained by the absence of an autocatalytic effect (O_2 does not influence the decomposition rate). In DINA decomposition (Fig. 10.6), the opposite situation applies due to the autocatalytic influence of nitrogen oxides. The upper curve in Fig. 11.18 (at $P_\infty = 1.6$ MPa) can be very well described by the theoretical relationship

Fig. 11.18 The Semenov diagram for thermal explosion of h/c H_2O_2 at various P_∞

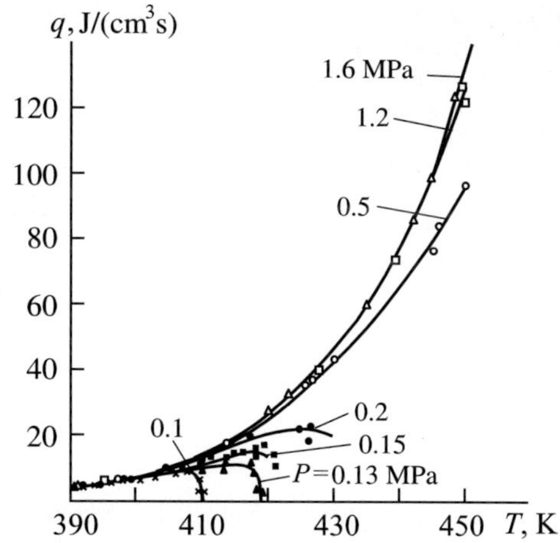

$$q = Qk_0 \exp\left(-\frac{E}{RT}\right), \tag{11.6}$$

where $Q = 2.75\,\mathrm{kJ\,g^{-1}}$, $E = 86.7\,\mathrm{kJ\,mol^{-1}}$ and $k_0 = 4.8 \times 10^5\,\mathrm{s^{-1}}$ (the values were obtained by kinetic methods, including low-temperature volumetric analysis).

Let us analyze the $q(T)$ dependencies obtained at low pressures, for example at $P_\infty = 10^{-1}$ MPa. In the low-temperature region the vapor pressure is significantly lower than P_∞, and the increase in the heat evolution rate as a function of the temperature can be described by Eq. (11.6). Then, according to relationship (10.17), the $q(T)$ function reaches its maximum value. At $T = 409$ K (the "adiabatic" temperature, T_{ad}) $q = 0$, which means that according to Eq. (10.11) the process becomes thermoneutral: the heat evolved upon decomposition of the condensed phase is completely consumed for the evaporation of the unreacted liquid. Due to the availability of reliable thermodynamic data for H_2O_2 [8] and its decomposition products, one can compare the experimental value for the "adiabatic" temperature $T_{ad} = 409$ K with that calculated using Eq. (10.11). For 96% (initial concentration) H_2O_2, the calculated value of the "adiabatic" temperature is equal to 410 K, which is in excellent agreement with the experimental value. According to Eq. (10.11), the "adiabatic" temperature depends on the process thermodynamics, and is not influenced by the kinetic parameters of the process. This is clearly illustrated by the experimental $q(T)$ dependencies (Fig. 11.19) obtained at $P_\infty = 10^{-1}$ MPa for two h/c H_2O_2 samples that differ in their homogeneous decomposition rates (TSI = 14 and 30): for both samples $T_{ad} = 409 \pm 1$ K (136 ± 1°C). Due to a relatively high value of the normalized heat-exchange parameter ($\alpha S/V \approx 8 \times 10^{-3}\,\mathrm{J\,cm^{-3}\,s^{-1}\,deg^{-1}}$), the dual-mode phenomenon was not observed at relatively low pressures ($P_\infty \leq 1.5 \times 10^{-1}$ MPa). In other words, a single value of the "stationary" warm-up ($\Delta T_{st} < \Delta T_{cr} = 17$–18 deg) applies to each value of T_0 up to $T_0 = T_{ad}$. The critical parameter value $\left(\frac{\alpha S}{V}\right)^{cr}_{lim} =$ $0.63 \times 10^{-3}\,\mathrm{J\,cm^{-3}s^{-1}\,deg^{-1}}$ at $P_\infty = 0.1$ MPa was calculated using expressions from

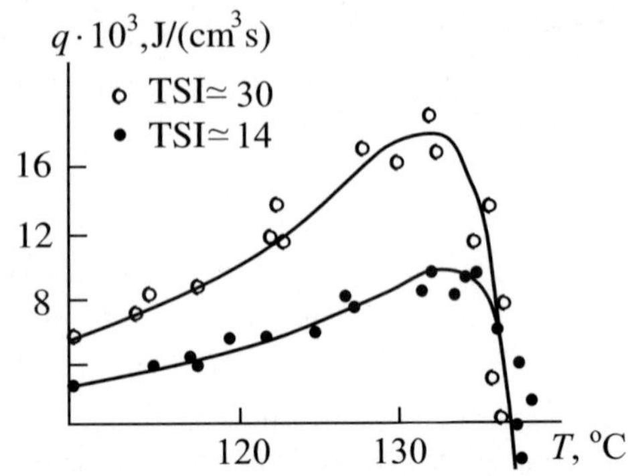

Fig. 11.19 Plot of the rate of heat evolution against the temperature for h/c H_2O_2 samples with TSI = 14 (*1*) and 30 (2) at $P_\infty = 0.1$ MPa

the theory presented in Sect. 10.4, kinetic data on homogeneous decomposition and the expression for the vapor pressure $P_{sv}^{H_2O_2} = 0.8 \times 10^5 \exp(-575/T)$MPa[8]. This value for the critical parameter is significantly lower than that indicated for the experimental setup used. At higher pressures ($P_\infty \geq 1$ MPa), both self-heating modes were observed, which is characteristic of regular (nonvolatile) systems prone to thermal explosion. However, in contrast to regular systems, the maximum value of h/c H_2O_2 warm-up measured with a thermocouple under above-critical conditions was quite high (3–5 times higher than that estimated using the Semenov theory: RT_0^2/E). In this case, the maximum liquid temperature was lower than the theoretical "adiabatic" temperature, T_{ad}. It was found from the thermograms that the decomposition of h/c H_2O_2 occurs in the flameless flash mode. After reaching its maximum ("non-Semenov") value, the warm-up rapidly drops. As the pressure increases further ($P_\infty \geq 1.5$ MPa), the warm-up of the liquid under above-critical conditions is accompanied by self-inflammation of the vapor. The process occurring under above-critical conditions is characterized by relatively poor reproducibility. This is because it is impossible to keep the conditions for heat and mass transfer in the gas phase above the vigorously decomposing liquid constant. Since the uniformity of the temperature in the liquid phase is ensured by its active mixing with rising oxygen bubbles, the approximation Bi $\ll 1$ used to calculate the critical conditions is quite correct. The corresponding experimental and theoretical data are in satisfactory agreement (Table 11.3).

Table 11.3 Comparison of theoretical and experimental parameters of the thermal explosion of h/c H_2O_2

$(\alpha S/V) \times 10^3$, J cm^{-3} s^{-1} deg^{-1}	P_∞, MPa	T_0^{cr}, K		ΔT_{cr}, K		ΔT_{st}, K	
		Theor.	Exp.	Theor.	Exp.	Theor.	Exp.
7.5	1.1	420	422	17.0	20.5	93	64
9.2	1.6	427	429	17.5	18.6	132	Vapor inflammation
12.9	1.7	433	434	18.0	19.0	140	Vapor inflammation

The difference between the experimental and theoretical values of the stationary above-critical (non-Semenov) warm-up can be explained by using a model that ignores the conversion degree. However, as was found from the above-critical thermograms, the conversion degree corresponding to ΔT_{st} is quite high ($\eta \sim 0.4$).

11.4 Thermal Decomposition of Hydrazine

The experimental determination of the kinetic parameters for hydrazine decomposition is quite difficult to achieve due to its low boiling point and its high (compared with h/c H_2O_2) thermal stability.

The composition of the gaseous products of commercial hydrazine decomposition was determined by a volumetric method using the AVD and a chromatograph. At 50–100°C and for low conversion degrees ($\eta \sim 1$–3%) the gas-phase products were found to consist of N_2 and H_2 in the ratio $N_2/H_2 \approx 18/20$ (within the experimental accuracy).

Noting that the ammonia produced completely dissolves in hydrazine, one can write the following equation for the overall process of the liquid-phase decomposition of hydrazine:

$$N_2H_4(l) \rightarrow \frac{4}{3}(1-x)NH_3(l) + \frac{1}{3}(1+2x)N_2(g) + 2xH_2(g), \qquad (11.7)$$

where $x \approx 10^{-2}$.

When the chemical analysis of the decomposition products was very difficult to perform (for manometric experiments at high pressures and thermographic experiments using hydrazine samples contained in sealed glass ampoules), they were assumed to consist of NH_3 and N_2 only. Taking into account the high thermal stability of hydrazine and its relatively low boiling point (at atmospheric pressure), and on the basis of estimates obtained using Eqs. (10.8a) and (10.8b), it was decided to run kinetic experiments at high temperatures and pressures. The behavior of hydrazine samples contained in sealed vessels (which prevented its evaporation) at $T \geq 373$ K was studied by manometric and calorimetric (of the thermographic type) methods. Thus the homogeneous decomposition rate of hydrazine was measured in the temperature range 318–532 K using the following techniques:

- *Volumetry* ($T = 318$–350 K, atmospheric pressure) using the Microdroplet device (I) and AVD (II).
- *Manometry* ($T = 350$–430 K, $P = 0.1$–5 MPa). The pressure as a function of time was measured by standard manometers installed on aluminum thermostated closed containers containing hydrazine (III). A method (developed by Prof. Nechiporenko) for measuring the pressure in a sealed hydrazine-containing vessel as a function of time was also used (IV).
- *Thermography* ($T = 535$–588 K, $P \leq 8$ MPa). The method used is similar to a technique described in Sect. 10.5.1, but in this case, hydrazine samples were

sealed in glass cylindrical ampoules (inner diameter 3.5–7 mm, wall thickness 1.5–2 mm, length 50 mm) placed in the electrically heated furnace of the thermograph (V).

Due to the high thermal stability of hydrazine, relatively high conversion degrees cannot be attained for the short durations of the low-temperature experiments (methods I–IV). The experiments were terminated after the process attained its stationary mode, characterized by a constant change in the volume (methods I–II) or pressure (methods III–VI).

In contrast, in the high-temperature (thermographic) experiments, hydrazine was found to be almost completely decomposed.

Typical thermograms for the high-temperature decomposition of hydrazine are shown in Fig. 11.20. The data were processed using the approach described above. The decomposition was found to occur according to a first-order reaction ($n = 0.90 \pm 0.05$, Fig. 11.21).

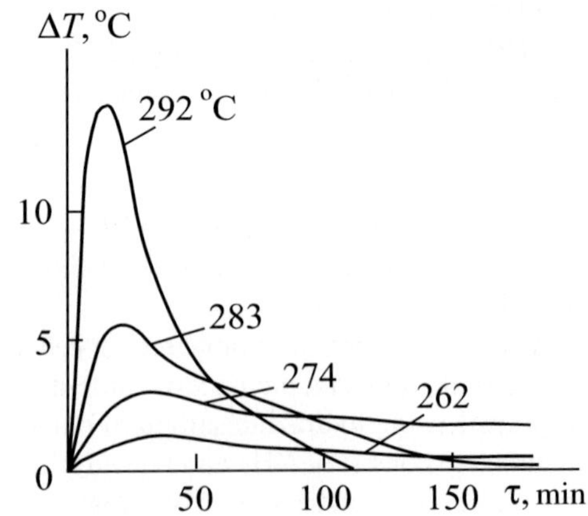

Fig. 11.20 Thermograms of the high-temperature decomposition of hydrazine at various temperatures

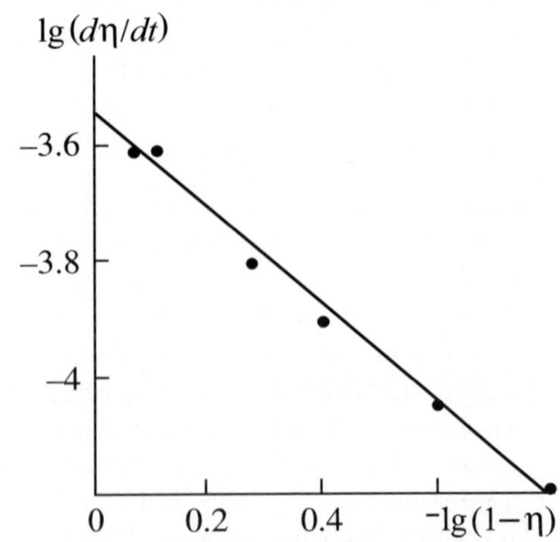

Fig. 11.21 Dependence of the isothermal rate of the high-temperature decomposition of hydrazine on the conversion degree at $T = 282$°C and $n \approx 0.86$

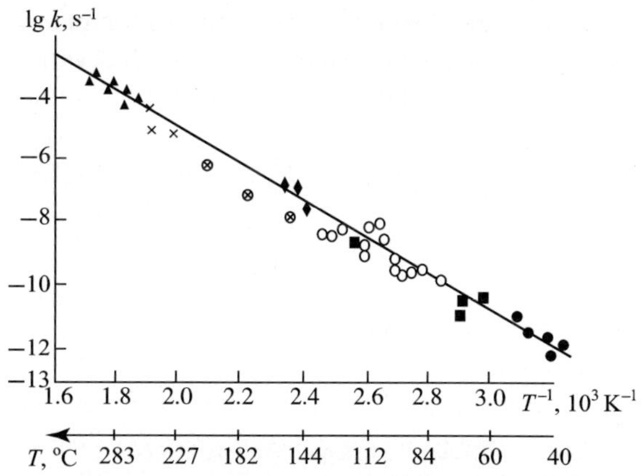

Fig. 11.22 Summarized data on the kinetics of the homogeneous decomposition of hydrazine in the liquid phase obtained by the Microdroplet method (*filled circles*), AVD method (*squares*), manometric method with standard manometers and aluminum containers (*unfilled circles*), manometric method developed by Prof. Nechiporenko (*diamonds*), thermography (*triangles*); data from [3] obtained using mass spectrometry (*crossed circles*), data from [4] obtained using the high-pressure manometric method (*crosses*)

A generalized Arrhenius plot including data obtained in this work as well as literature data from [3] (obtained by applying mass spectrometry to the products) and [4] (high-pressure manometry using differential pressure mercury gauge) is shown in Fig. 11.22.

The kinetics of the homogeneous decomposition of hydrazine does not change over a wide temperature interval ($\approx$ 240 deg) and an extremely wide range of the rate constant ($\approx$ 9 orders of magnitude). The expression for the rate constant is

$$k = 5 \times 10^{6.0 \pm 0.2} \exp\left(-\frac{26800 \pm 800}{RT}\right) [s^{-1}]. \quad (11.8)$$

An express method for evaluating the stability of hydrazine (using the Microdroplet device) under long-term storage conditions was developed based on the kinetic data

Table 11.4 Kinetic constants for the heterogeneous decomposition of hydrazine on the surfaces of construction materials

Material	Temperature, °C	Preexponential factor k_{0S}, cm s^{-1}	Activation energy E, kJ mol^{-1}
Stainless steel H18N10T	30–260	13	66.7
Stainless steel 12H18N10T	40–130	380	76.7
Nickel-based alloy EI654	40–130	15	72.9
Stainless steel 12H21N51	40–130	29	72.1
Teflon film FT-4MB	50–95	25	82.9
Aluminum-based alloy AMG-6	50–130	12	90.8

obtained [17]. This method is used in one of the chemical plants in Russia for quality control of commercial hydrazine (the maximum gas evolution is 1–2 l/ton during a year of storage at 20°C) and bench tests (the maximum gas evolution is higher than that indicated above).

The kinetics of the heterogeneous decomposition of hydrazine on surfaces of construction materials were studied using the approach described in Sect. 11.3.2. The data obtained are summarized in Table 11.4.

References

1. Ivanov DV (1977) At the entrance to Mirnyy. Plesetsk—Moscow. Gamma, Moscow
2. Borenstein S (2003) The Mercury News, Feb 1
3. Davis AS (1979) AIAA Paper 90–1020
4. Munik GS (1992) Pure Appl Chem 64:965
5. Eberstein JJ, Glassman J (1965) 10th Symp on Combustion, Cambridge, UK, 17–21 Aug 1964
6. Genich AP, Zhirnov AA, Manelis GB (1974) Rus J Phys Chem 48:416
7. Alexeev AP, Manelis GB (1977) Combust Explos Shock Waves 16:54
8. Shamb U (1958) Hydrogen peroxide. Inostrannaya Literatura, Moscow
9. Haber F, Weiss J (1934) Proc Roy Soc 147:332
10. Emmanuel NM, Denisov ET (1952) Dokl Akad Nauk SSSR 83:594
11. Emmanuel NM, Denisov ET, Mayzus ZK (1965) Chain reactions of liquid-phase hydrocarbon oxidation. Nauka, Moscow
12. Nadezhdin AD, Pourmal' AP (1973) Kinet Katal 14:141
13. Shtamm EV, Pourmal' AP, Skurlatov YuI (1991) Rus Chem Rev 60:1228
14. Brounshtein BI, Denisov ET, Slutsker BM, Shteinberg AS (1971) Kinet Katal 5:213
15. Muratov SM, Makharinsky VM, Afanas'ev GT (1972) In: Stesik LN (ed) Combustion and explosion. Nauka, Moscow, p 30
16. Shteinberg AS, Volodina LE (1990) Method for evaluation of thermal stability of explosives. State Institute of Applied Chemistry, St. Petersburg, Russia
17. Price EW (1993) In: 30th JANNAF Combustion Subcommittee Meeting, Monterey, CA, USA, 15–19 Nov 1993

Final Conclusions

Here we review the most important results from the studies presented in this book.

1. The qualitative aspects of the macrokinetics of the fast high-temperature decomposition of condensed energetic materials, including explosives, gunpowders and the majority of modern propellants (solid, liquid and hybrid), were thoroughly analyzed. It was shown that phenomena resulting from the spatiotemporally nonisothermal character of the processes and equilibrium physicochemical transformations (evaporation, sublimation and dissociation) play a very important role. Therefore special experimental and analytical methods for studying these processes need to be developed.
2. A theory of the linear pyrolysis of condensed compounds was developed. Analytical relationships between experimentally measured characteristics (linear decomposition rate, temperature of the pyrolysis surface, parameters of outward heat exchange, etc.) and the kinetic constants of the limiting decomposition reaction were derived. Depending on the parameters, linear pyrolysis was found to occur in two modes: kinetic (the pyrolysis macrokinetics are identical to the true decomposition kinetics and $E_{\text{eff}} \approx E_{\text{true}}$) and inner diffusion ($E_{\text{eff}} \approx E_{\text{true}}/2$).
3. Devices for measuring the basic characteristics of the linear pyrolysis of components of composite solid and hybrid propellants (linear decomposition rate, temperature of the pyrolysis surface, total gasification heat, and others) at high pressures and in vacuum were designed. The following unique devices can be used to obtain data that were previously unavailable: (i) "chemical arc," for measuring the rate of pyrolysis due to the interaction flame for oxidizers and fuels of composite solid propellants and for identifying the area of action of the combustion catalyst; (ii) a device for performing the linear pyrolysis of thin films with double thermostating, for direct determining the decomposition mechanism (volumetric or surface); (iii) block calorimeter, for the express determination of the total heat values for the high-temperature pyrolysis of condensed compounds, including thermoplastic combustible composite solid and hybrid fuels.
4. Linear pyrolysis of model binders (linear and crosslinked PMMA) was studied over wide parameter ranges. The experimental dependencies of the pyrolysis

rate on the surface temperature were found to be in qualitative agreement with conclusions made based on the theory of linear pyrolysis. The kinetic parameters of high-temperature PMMA decomposition estimated using the formula from the theory were shown to be in good quantitative agreement with kinetic data on the volumetric decomposition of PMMA. A new "third" mode of high-temperature linear pyrolysis of polymers was described for the first time. This mode is characterized by an active increase in the pyrolysis rate as the thermal flux rate increases for insignificant growth of the surface temperature after it reaches a certain (polymer-specific) value. The third mode needs further studies, but it is clear that it is significantly influenced by a change in the kinetic mechanism (from "end initiation" to "random chain rupture"), the evaporation of large chain fragments and the dispersion of unreacted polymer particles.

5. At standard rates of burning in combustion chambers of hybrid rocket engines, the pyrolysis of hybrid polymer-based propellants was found to occur in the third mode. This explains the weak influence of the kinetic constants of the high-temperature decomposition of polymers on the burning rate. The universal character of the corresponding dependence was proven by comparative analysis of the experimental total heat values for the high-temperature gasification of a number of model and commercial hybrid propellants and the rates of burning of these fuels in hybrid rocket engines. In other words, the total heat of gasification is the main characteristic of the high-temperature decomposition of the material, which determines its rate of linear burning in a hybrid rocket engine. An express method for the preliminary evaluation of the hybrid propellant burning rate from data obtained using high-temperature calorimetry was developed.
6. A simple and efficient method for evaluating kinetic constants for the initial stages of the high-temperature decompositions of homogeneous blasting powders and heterogeneous composite solid propellants from their ignition characteristics using a hot heat-conducting block was developed. The kinetic patterns of the corresponding processes for some practically interesting homogeneous materials (nitrocellulose, organic azide APU, and others) and a wide range of AP-based model and commercial composite solid propellants were studied. Expressions for the kinetic constants for limiting decomposition reactions obtained from data on the ignition of composite solid propellants were found to differ significantly from those for low-temperature decomposition. It was shown for commercial polyurethane-based composite solid propellant (used as an example) that processes determining the kinetics of the high-temperature decompositions of AP–polymer mixtures also occur at relatively low temperatures. However, in the latter case they are not very apparent due to the very low reaction rates caused by very high activation energies. The parameter ranges within which the high-temperature decomposition of composite solid propellants accompanying ignition and burning can be treated as being homogeneous were determined. Aluminum was found to have no influence on the high-temperature decompositions of composite solid propellants.
7. DTA- and TGA-based methods were used to obtain data on the kinetics of the fast high-temperature decomposition of propellants. Two approaches to deep reagent ballasting with an inert compound ("mechanical" and "thermal"

dilution) that are used for homogeneous and heterogeneous systems were proposed. These techniques allow valuable information to be obtained about the kinetics of fast deep transformations in condensed systems at high temperatures.

8. Automated devices used to obtain data on the rate of heat evolution under isothermal conditions and programmed heating (at pressures from 10^{-3} to 5 MPa and temperatures from 300 to 800 K) were designed, manufactured and applied for laboratory studies. Methods and volumetric devices for measuring gas evolution rates covering seven orders of magnitude (10^{-6} to $10\,\mathrm{ncm^3\,s^{-1}}$) were developed.
9. The macrokinetics of AP (the most commonly used oxidizer for modern composite solid propellants) decomposition was quantitatively characterized over a wide range of temperatures and pressures by the linear pyrolysis method and thermography. The decomposition kinetics was shown to be constant over a range of rate constants covering six orders of magnitude. The process was found to occur via two stages: the fast decomposition of a small amount of the reagent (12–15%) and the relatively slow decomposition of the rest of the compound. The kinetic parameters of the stages were determined, and a generalized equation describing the high-temperature decomposition of AP was obtained. A specific mode of AP linear pyrolysis accompanied by the burning of gaseous products of dissociative sublimation at $P \approx 0.1$ MPa and $T_S \approx 830$ K was observed. The parameters of this mode and the experimental dependence of the total thermal effect of the high-temperature decomposition of AP on the pressure were explained in relation to irreversible decomposition and reversible processes that occur simultaneously in volatile explosives.
10. The linear pyrolysis of components of a model composite solid propellant (AP and PMMA) was studied under the conditions of burning in the chemical arc. It was shown that two flames coexist in the gas gap separating the components: a deflagration flame near the AP surface and an interaction flame near the PMMA pyrolysis surface. Burning in the interaction flame at 0.1 MPa occurs with an oxidizer deficiency. The effect of introducing the combustion catalysts CuO and Fe_2O_3 separately into fuel and oxidizer on linear pyrolysis in the chemical arc was studied. Fe_2O_3 was found to influence the process when it was present in the polymer, while the effect of CuO could be seen when it was introduced into the oxidizer. This was explained on the basis of ideas about how the mechanism of the catalyst influences the kinetics of AP thermal decomposition and the conversion of hydrocarbons. It was shown that the chemical arc-based technique is quite promising for studying the mechanisms of pyrolysis and combustion of composite solid propellant components.
11. High-temperature decomposition of AP-based model mixtures and commercial composite solid propellants was studied using thermography and thermogravimetry over wide ranges of temperatures, pressures and conversion degrees. For almost all of the studied materials, the overall process was found to occur via two stages characterized by activation energies of 125 and 229–250 $\mathrm{kJ\,mol^{-1}}$. For mixtures based on polymers that have different structures and thermodestruction characteristics, the first-stage rate constants differ significantly, while the second-stage kinetic parameters are almost identical. A strong inhibitory

influence of the pressure on the kinetics of the second stage of the high-temperature decompositions of AP-based composite solid propellants was observed and studied. These phenomena were qualitatively explained by the staged nature of the polymer oxidation and by the dependence of the relative fraction of the equilibrium AP dissociation product (chloric acid vapor) in the gas oxidizer produced by AP decomposition on the ambient pressure.

12. The decomposition macrokinetics and thermal explosion of volatile explosives and monopropellants were analyzed. In contrast to nonvolatile explosives and gas systems, self-heating and especially thermal explosion was found to be complicated due to the absorption of heat consumed during phase transitions (evaporation, sublimation) in the volume. The ambient pressure was shown to affect the critical condition due to the specific characteristics of the dependence of the rate of heat evolution on the temperature and ambient pressure. New modes above the explosion initiation limit that are characterized by quasi-stationary warm-ups that significantly exceed the maximum stationary warm-up value from Semenov theory were theoretically predicted and analyzed. Expressions for quantitatively evaluating the contribution of the volatility to the reaction macrokinetics in calorimetric and thermogravimetric kinetic experiments were obtained. An analytical scheme for determining the critical conditions for thermal explosion of volatile explosives and monopropellants was proposed. The critical condition was found to be determined by the kinetics of the gas-phase reactions. The process associated with the region above the explosion initiation limit can occur in two modes: as a vigorous flameless flash or as a process of vapor self-inflammation, depending on the macrokinetic conditions of the vapor-phase reaction. The correctness of the theory developed was quantitatively verified by experimental data on the thermal explosions of DINA (volatile explosive) and h/c H_2O_2 (monopropellant). Contradictions between the explanation for the inflammation mechanism of these compounds proposed by Belyaev (flash in the vapor phase) and the experimental results obtained by Merzhanov (the critical condition for thermal explosion is determined by the reaction kinetics in the condensed phase) were also resolved.
13. The homogeneous decomposition of the promising monopropellants h/c H_2O_2 and N_2H_4 were studied over a wide temperature range using modern physical methods including newly developed ones (described in this book). The expressions for the rate constants were obtained for H_2O_2 over the temperature range of 300–445 K and a reaction rate interval covering 5.5 orders of magnitude, and for N_2H_4 over the temperature range of 310–580 K and a reaction rate interval covering nine orders of magnitude. The proposed express methods for evaluating the stabilities of these monopropellants were implemented for laboratory research and factory tests.
14. The kinetics of the heterogeneous decompositions of h/c H_2O_2 and N_2H_4 on the surfaces of the most commonly used construction materials was studied. The data obtained have been successfully used in R&D laboratories and engineering and design departments to evaluate the stabilities of these monopropellants and develop safety recommendations for their storage and application in rocket and space engineering and technology.

Index

Printing: Krips bv, Meppel, The Netherlands
Binding: Stürtz, Würzburg, Germany